The authors are barristers at Temple Gard

Alex Glassbrook is a specialist in the law vehicles. He wrote the first British book on the subject, "The Law of Driverless Cars: An Introduction" (also published by Law Brief Publishing) in 2017. He contributed to the civil liability part of the Law Commission's 2018 consultation on the law of automated vehicles. His practice features all aspects of civil road transport law, including motor insurance, counter-fraud and catastrophic personal injury cases. Alex is recommended as a leading barrister by both of the British bar directories (Chambers UK Bar and the Legal 500), which have described him as "skilled in handling digital evidence and technology development issues", "a true expert" and "the most forensic advocate".

Emma Northey specialises in personal injury, inquests, civil fraud and regulatory law. Before coming to the Bar, Emma was a Trading Standards Officer for 15 years. She has an in-depth knowledge of all aspects of consumer protection and regulatory crime, and a particular interest in the challenges posed by the regulation of novel goods and services. Emma's interest in inquest law began during her time as an investigator, dealing with the tragic consequences of the supply of defective products. More recently, she has acted for Interested Persons in a number of complex and high profile inquests.

Scarlett Milligan has a busy practice encompassing public law, regulatory work and personal injury claims. This gives her a valuable insight into the numerous issues which driverless vehicles are likely to present at the policy level, in practice, and across the broader legal landscape. She has a particular interest in the civil regulation of human activity by the state, and the policy choices which have to be made in the management of risk within society. A primary focus of her work on the law of driverless vehicles has been the novel duties which may be imposed – whether by the courts or legislation – in respect of the manufacture and operation of driverless vehicles, and the challenges associated with such duties coming into operation alongside existing regimes. Scarlett looks forward to seeing how the courts grapple with these issues as well as the problems posed by human interaction with robotics and artificial intelligence systems.

A Practical Guide to the Law of Driverless Cars
Second Edition

A Practical Guide to the Law of Driverless Cars
Second Edition

Alex Glassbrook, B.A. (Hons)
Barrister of the Middle Temple

Emma Northey, B.Sc. (Hons), DTS
Barrister of Lincoln's Inn

Scarlett Milligan, LLB
Barrister of the Inner Temple

Law Brief Publishing

© Alex Glassbrook, Emma Northey, Scarlett Milligan

All rights reserved. No part of this publication may be reproduced, stored in a retrieval system, or transmitted, in any form or by any means, electronic, mechanical, photocopying, recording or otherwise, without the prior permission of the publisher.

Excerpts from judgments and statutes are Crown copyright. Any Crown Copyright material is reproduced with the permission of the Controller of OPSI and the Queen's Printer for Scotland. Some quotations may be licensed under the terms of the Open Government Licence (http://www.nationalarchives.gov.uk/doc/open-government-licence/version/3).

Cover image © iStockphoto.com/metamorworks

The information in this book was believed to be correct at the time of writing. All content is for information purposes only and is not intended as legal advice. No liability is accepted by either the publisher or author for any errors or omissions (whether negligent or not) that it may contain. Professional advice should always be obtained before applying any information to particular circumstances.

Published 2019 by Law Brief Publishing, an imprint of Law Brief Publishing Ltd
30 The Parks
Minehead
Somerset
TA24 8BT

www.lawbriefpublishing.com

Paperback: 978-1-912687-68-8
Kindle ebook: 978-1-912687-69-5

PREFACE

Thank you for taking an interest in Driverless Law, and particularly the British laws of driverless vehicles. Neither term is quite correct, but we will deal with that shortly.

It is an unusual – and exhilarating – time to be writing about the law in Britain. On 24 January 2017, the day Alex wrote the Introduction to the first edition of this book ("The Law of Driverless Cars: An Introduction"), the Supreme Court had just handed down its judgment in what we must now call the first Gina Miller case, which engaged questions of, even then, rare constitutional importance.

The fact that the second major constitutional judgment of very many years[1] was handed down only a week ago, while we were writing the final chapters of this second edition, seems even more remarkable.

Happily – and despite the impression you might have had – there have been other innovations in British law, apart from the constitutional, since January 2017. And some of them have been in Driverless Law.

The first was the appearance on the statute book of the Automated and Electric Vehicles Act 2018, on 19 July 2018. Part Two of this book deals with that Act (the AEVA) in detail.

The birth of the AEVA was a tribute to all the policymakers involved – perhaps especially all those at the Department of Transport's Centre for Connected and Autonomous Vehicles (the CCAV) who had worked so hard on the Bill. That the UK was among the few nations so far to have legislated for CAVs was a conspicuous achievement.

The second was the launch of the joint consultation, by the Law Commission of Scotland and the Law Commission of England and Wales, on the law of automated vehicles. The joint consultation paper (to which we refer, with links to the Law Commissions' Automated Vehicles project

---

1 *R (on the application of Miller) v The Prime Minister* [2019] UKSC 41, 24 September 2019.

website[2], in this book) was a masterwork: a model of how to approach a factually and legally difficult subject with an intense commitment both to scholarship and to practicality. That so many engineers, insurers, public authorities, academics, lawyers and others responded so energetically is a tribute to Jessica Uguccioni, the lead lawyer on the Law Commissions' AV project, and her brilliant and industrious team. The Law Commissions' sponsorship of the Model Law Commission – a project enticing state school students to the law, started by students at City University – and in particular of the Model Law Commission's own AV project, produced an extremely valuable presentation by the volunteer students in parliament. As a state school and City University alumnus, I was especially thrilled to be involved in that project.

The third factor – and the very best piece of luck – was that I (Alex) was joined in this second edition by two co-authors of such quality. Emma Northey has been my colleague at Temple Garden Chambers for several years, and is an extraordinarily able, calm and thoughtful lawyer: the sort of opponent whom you wish for in any case for their intelligence and company, but also dread for their quality as opposing counsel. Scarlett Milligan had the misfortune to be my pupil, but was clearly going to succeed despite that disadvantage. Scarlett's capacity for hard work and legal insight when writing this book, against the demands of her very busy practice, has been typical of her but remarkable nonetheless. Collaborating with Emma and Scarlett could not have been a more enjoyable and productive experience. This book would have been a lesser enterprise, in every way, without them. I am extremely grateful to them both for giving it their hard work and expertise.

Finally, we would all like to thank Tim and Garry at Law Brief Publishing for their patience, their unfailingly cheerful support and for asking us to write this book in the first place. We share their enthusiasm, but any errors are our own.

Alex Glassbrook
Emma Northey
Scarlett Milligan
Temple Garden Chambers, London
1 October 2019

---

2 https://www.lawcom.gov.uk/project/automated-vehicles/

CONTENTS

Introduction		1
PART ONE	**WHERE WE ARE NOW** *Scarlett Milligan*	11
Chapter One	Life before the Automated and Electric Vehicles Act 2018	13
PART TWO	**THE AUTOMATED AND ELECTRIC VEHICLES ACT 2018** *Alex Glassbrook*	31
Chapter Two	Overview of the AEVA 2018	33
Chapter Three	Causation in the AEVA 2018	49
Chapter Four	Contributory Negligence in the AEVA 2018	59
Chapter Five	Exclusions and Restrictions in the AEVA 2018	71
Chapter Six	The Insurer or Owner's Right to Claim Against Another Responsible Person Under the AEVA 2018	85
PART THREE	**AUTOMATED VEHICLE LAW ISSUES OUTSIDE THE AEVA 2018**	93
Chapter Seven	The New Regulators *Emma Northey*	95
Chapter Eight	Uninsured Vehicles *Emma Northey*	111

Chapter Nine	Product Liability Claims *Scarlett Milligan*	119
Chapter Ten	Data and Privacy *Alex Glassbrook*	153
Chapter Eleven	Criminality *Emma Northey*	177
Chapter Twelve	Employers Liability Claims *Scarlett Milligan*	185
Chapter Thirteen	Equality *Alex Glassbrook*	217

INTRODUCTION

OBJECTIVES OF THIS BOOK

It might not seem an obvious time to write about British Driverless Law. We are, in one sense, awaiting the next bus: the AEVA 2018 is on the statute book but Part 1 of the Act (dealing with automated vehicles – the term which the Act gives to autonomous vehicles) has yet to be brought into force, by regulations to be made by the Secretary of State for Transport.

However, in our view it is a very good time to take stock of the law.

In the first place, the British law of driverless vehicles is relatively advanced, in that there is an Act in place, which sets a particular course[1] and is ready to be activated.

Second, while technology is unpredictable, the signs are that automated vehicles, at SAE Level 3 (see below), might over the next few years reach the market. There is good reason to see that period as the likely growth period for electric vehicles (EVs). With a growth of EVs, automated features seem likely to come into greater use.

Third, the legal questions behind the current law of AVs are truly difficult. Questions of causation of an accident become particularly troublesome when the main tortfeasor is no longer human but machine. Questions of contribution between responsible parties complicate when the machine is not only hardware but software.

Our laws are not yet accustomed to those questions, nor the many others which connected and autonomous vehicles provoke (including problems of insurance, employment, data and privacy and of equality, topics which we discuss in the third part of this book). Society might

---

1 Part 1 of the Automated and Electric Vehicles Act 2018 sets the course for at least part of the law of future road transport, namely civil liability for accidents caused by AVs, and the insurance foundation upon which it rests: see Part 2 of this book.

need to answer those questions soon. It cannot be too early to start thinking.

THE LANGUAGE OF "DRIVERLESS"

Like its predecessor (Alex's 2017 book: "The Law of Driverless Cars: An Introduction") this book uses the word "driverless" in its title.

That is shorthand. There is both a lively debate as to the proper use of technical terms, and a large quantity of those terms.

The point is well-made that lawyers and engineers should speak the same language[2] in order to understand autonomous vehicle technology and its implications for law and policy.

We agree with that argument and have tried to follow it in our preparation of this book. However, we are not engineers so, while applying ourselves as carefully as we can to the engineering information, we will try not to hazard opinions outside our legal expertise. Any errors in relation to the engineering facts or otherwise are of course our own. If we wander from the path in this edition, we would welcome warning shouts from our readers, so we can stick to the preferred route in the next edition.

We apologise in advance for using terms (eg. "driverless") which might appear too general to a technical reader[3]. We are aware that this might irritate some readers, who are already familiar with the lexicon. Different writers (and organisations) use different terms. So some compromise seems better, to keep the language simple and accessible to all.

---

2 Bryant Walker Smith, "Lawyers and engineers should speak the same robot language", Chapter 4 of "Robot Law" (Edward Elgar Publishing, 2016)

3 See eg. Section 7, page 28 of the SAE definitions J3016, June 2018 (see footnote below): "Deprecated Terms". Our approach recognises the importance of understanding the correct terms but differs in expression: many terms are in common usage, so we try to explain, not to exclude, those terms.

In a subject full of technical terms and abbreviations, we try to use as few as we can. The technical detail of the subject is deep and getting deeper. But, with the growth of such terms, there is a danger of language obscuring its own meaning.

With apology, there are some abbreviations which are helpful to keep in mind. This term appears a lot: CAV (Connected and Autonomous Vehicle).

Within that term, "CAV", there are two types of feature:

- A Connected Vehicle ("CV") is a vehicle fitted with communications devices which provide information to the driver or to the vehicle, allowing either to collaborate with other vehicles (abbreviated to "V to V" connectivity, or "V2V"), with parts of the road infrastructure ("V to I" or "V2I") and even with other objects such as bicycles or pushchairs, where the user has a connected device, and similarly-equipped pedestrians ("V to X" or "V2X").

- An Autonomous Vehicle ("AV") is a vehicle designed to be capable of safely completing journeys without the need for a driver, in all normally encountered traffic, road and weather conditions[4].

Mobile communications technology has been perhaps the single greatest driver of technological change over recent decades. Road transport technology develops in and because of that environment. So it seems overwhelmingly likely that a driverless vehicle will be driverless because it has both autonomous capability (its own capability to sense its surroundings and learn from them – by what has been described as "machine learning" – via its own hardware and software) and because it

---

[4] This would be the pinnacle of AV technology – the vehicle which the reader is likely to think of as truly "driverless". In fact, there are gradations of AV: see the discussion of the levels of automation, below.

is connected to other things, from which to learn (ie. it can communicate with other road users and infrastructure).

So "Connected and Autonomous Vehicle" (CAV) is the umbrella term, because it describes the twin features of road vehicles of the future.

We raise this point in the introduction, because the term "Connected and Autonomous Vehicles" also provides a useful prism through which to view legal issues.

Some legal consequences (for example, in relation to data and privacy) are likely to arise from the connections between vehicles, and from the data which those vehicles share with other systems (so, occasionally, the term "CAV" has a particular significance - for example in Chapter 10, "Data and Privacy").

Other legal consequences (such as liability to compensate victims of any road traffic accidents involving CAVs) are likely to flow from the autonomous side of the technology, and to involve questions about the attribution of blame between the automated systems of the car on the one hand and the human actors (driver, pedestrian, cyclist et al) on the other hand.

THE CLASSIFICATION OF EXISTING VEHICLES AND CAVs

The fully-autonomous vehicle has not yet been released for sale (nor built, so far as we know). But the vehicles of the future have already been classified.

The classification system has been refined, as more details emerge of manufacturers' plans for CAVs.

The essential question for many readers of the classification is: "Will I still need to drive the car?". That is also currently the central question relating to human liability for accidents and criminal offences of driving.

INTRODUCTION • 5

There are several variations, but the essence of the question is: "When will I be able to leave driving entirely to the car, and go to sleep/read a book/watch a film?".

The only clear answer is "not yet". But when we will reach that point is unclear. In the classification, it is unsurprisingly at the end point: Level 5. But whether the human driver might pass responsibility to the machine (and liability to an insurer) at an earlier stage is an open question.

For the sake of clarity, we distinguish between the "human driver" – the current, invariable norm – and the future "robot driver". By that we do not mean that an android[5] will be in the driver's seat but that the car itself will be robotic: "a machine capable of carrying out a complex series of actions automatically"[6].

The core of the classification system is the distinction between 6 levels of vehicle automation, from 0 to 5. The full details of the classification have been much discussed and the most authoritative standard, the Society of Automotive Engineers' or "SAE" standard[7], was revised in June 2018. We would summarise the 6 levels in the SAE standard as follows:

---

[5] "(In scienace fiction) a robot with a human appearance": Oxford English Dictionary (OED)

[6] OED

[7] The key source for the classification of levels of driving automation is the Society of Automotive Engineers' "Taxonomy and Definitions for Terms Related to Driving Automation Systems for On-Road Motor Vehicles" (which we abbreviate to "the SAE definitions"). The current version of the SAE definitions (the June 2018 revision) is at https://www.sae.org/standards/content/j3016_201806/. Please note that access to this document requires registration on the SAE site. And an updated version of the SAE graphic showing the SAE levels of automation was published in January 2019: https://www.sae.org/news/2019/01/sae-updates-j3016-automated-driving-graphic

Figure 1: our summary of the SAE levels of driving automation.

Level Number	Label	Who is driving? Human and/or Robot?	Description
0	No driving automation	Human	Conventional vehicle
1	Driver assistance	Human	Conventional vehicle with some assistance functions such as cruise control
2	Partial driving automation	Human and Robot, with Human supervising and completing dynamic driving tasks started by Robot	Conventional vehicle with advanced driver assistance functions eg. lane-keeping and collision avoidance systems
3	Conditional driving automation	Robot and Human, with Robot carrying out driving for most of the time in circumstances familiar to it[8], but with Human available to intervene when appropriate, to bring the vehicle to a safe state[9].	Automated Driving System (ADS), in which majority of driving is automated, with human ready to intervene if appropriate

---

8   "Circumstances familiar to" a connected or autonomous vehicle is our précis of the technical phrase "Operational Design Domain" (ODD), meaning the environment in which an Advanced Driver System (ADS) is capable of operating, by design. See page 3 of the SAE Taxonomy for more detail.

INTRODUCTION • 7

4	High driving automation	Robot driving in circumstances familiar to it, with human supervision but without expectation of human driver needing to intervene in those circumstances	High ADS: almost all driving automated, with heavy presumption that robot does all driving for which it is designed, but requiring residual human supervision
5	Full driving automation	Robot has control of car in all circumstances	Full ADS, no human driver required: **Driverless**

Beneath the thicker black line, from levels 3 to 5, are the systems referred to collectively as "Automated Driving Systems" (ADS), where the balance of the greater part of the driving task switches from Human (at level 2) to Robot (at level 3) then progresses towards the entire automation of driving (at level 5).

At the time of publication, in late 2019, mainstream new car technology is at level 2. Manufacturers are in a race to release tested Advanced Driving Systems into the market at Level 3 and beyond.

---

9 In the technical language of the SAE, a safe state for the vehicle (for example, brought at a halt in a safe place at the side of the road) is referred to as the "minimal risk condition". The driver ready to respond to a request by an Advanced Driver System to intervene, to bring the vehicle back to a minimal risk condition, is known as the "fallback-ready user". The fallback-ready user is a feature unique to Level 3 vehicles in the SAE standard: Levels 0 to 2 (current, conventional vehicles) all require a human driver to be in charge, and in Levels 4 and 5 it is presumed that the robot system will be ready and able to intervene to take the vehicle back to the safe place. The concept of Fallback Ready User is practically – and legally – problematic, as it raises difficult questions of when and how the human user should resume control from the robot system. The Law Commission describes Level 3 as occupying the "murky middle" of AV liability law.

The SAE level of automation at which manufacturers first release CAVs onto the market will be significant: Level 3 (at which both human driver and robot system share control of the vehicle) raises difficult questions as to when and how the human driver should intervene to take back control of a moving CAV. The insurance industry is wary of the risks of Level 3, and the British Government has stated that the strict liability-type provisions of the Automated and Electric Vehicles Act 2018 will not apply to Level 3, but only to Level 4 and 5 CAVs (see Chapters 2 and 5).

A NOTE ON LAW AND JURISDICTION

We are British lawyers, practising in the United Kingdom (and, as barristers, we are members of the Bar of England and Wales). So, we approach these questions from a common law perspective and with the law of England and Wales as our focus.

However, technology knows few (if any) boundaries, and we have some acquaintance with laws beyond the borders of the UK, particularly European Union (EU) law and the laws of the United States. We would be happy to explore other legal perspectives in future editions of this book (and, again, would welcome any relevant comments on this edition).

The law is the law of England and Wales, as at 1 October 2019. This book should not be relied upon as a comprehensive account of the law of and relating to automated vehicles from that date, without further research by the reader, because that law has yet to take a comprehensive form. In particular, regulations will (at an as yet unfixed date) be made by the Secretary of State both to bring Part 1 of the AEVA 2018 into force and to shape regulation of automated vehicles.

This book is not the setting in which to describe all the detail of particular legal topics relevant to its subject of CAV law (eg. equality, employment, motor insurance etc). Those topics are large and cases turn upon their particular facts. For such further detail the reader is referred to specialist works.

A NOTE ON FOOTNOTES, SOURCES AND WEBSITES

We cite sources in footnotes. Several of those sources are online. Please note that, though we take care to cite only those sources which we regard as authoritative, we are unable to give any guarantee as to service on websites, nor any guarantee as to the accuracy of information.

CAV technology is fast-developing and the source of a great deal of news and speculation, so we have tried to strike a balance between keeping the reader informed of what we see as the main points, and avoiding over-reliance upon fast-changing comment.

THE END OF THE BEGINNING…

This book is unavoidably an exercise in speculation. We cannot yet predict legal results in CAV cases with confidence, any more than we can predict how the technology will evolve.

But the mist is starting to lift from the new landscape, the shapes of unfamiliar technologies are emerging and we can start to glimpse the way ahead.

We hope that this book will provide you with a map of at least the major features of that landscape – and of some of the intriguing byways.

PART ONE

WHERE WE ARE NOW

SCARLETT MILLIGAN

CHAPTER ONE
LIFE BEFORE THE AUTOMATED AND ELECTRIC VEHICLES ACT 2018

OBJECTIVES OF THIS CHAPTER

The objectives of this chapter are to:

- Outline key aspects of the CAV technology which is likely to appear on our roads in the next few years;

- Briefly explore the role of negligence in road traffic accidents today;

- Explain the new and interesting problems which CAV technology will pose to the current negligence framework;

- Highlight likely points of dispute between parties in negligence cases involving CAVs, and discuss how our laws may need to evolve to fully accommodate and resolve such disputes;

- Explain how the Automated and Electric Vehicles Act 2018 ("the AEVA") may (or may not) resolve the problems highlighted.

THE EARLY EMERGENCE OF CAV TECHNOLOGY ON OUR ROADS (SAE LEVELS 1-3)

Any book or discussion exploring artificial intelligence ("AI") and Connected and Autonomous Vehicle ("CAV") technology is likely to strike a reader as not only complex, but somewhat futuristic: after all, fully autonomous cars, taxis and delivery vehicles are not expected to arrive on our roads in the immediate future. Moreover, even when those vehicles do arrive, it will take some years before they form the *majority* on our roads and can truly be described as embedded in our daily lives. Thus, many readers may be wondering: it is too early to contemplate the impact of CAVs on road traffic accident law? Surely the future arrival of CAV technology will be accompanied by a specific legal framework?

The reality is that elements of CAV technology will be with us sooner than one might expect; indeed, readers will already be familiar with some aspects of automated technology, for example: cruise control; lane-keeping tools; automatic parking, or parking assist systems. These technologies fall within SAE Levels 1 and 2, and are already beginning to give rise to legal and technical questions and problems, such as:

- Will a lane-keeping tool be programmed to move a car to accommodate or avoid a cyclist? Or will the car need to be programmed to maintain its position to prevent it colliding with other traffic? Would the cyclist or the vehicle be responsible for a collision in such circumstances? How *should* the vehicle have been programmed?

- If I ask my car to park itself in a narrow space, am I to blame if another vehicle is damaged, or is it the fault of the vehicle for failing to park correctly, or to otherwise say 'no'?

Whilst some manufacturers are focussing on moving straight from existing technologies to Level 5 (i.e. "fully autonomous") vehicles, others appear to be phasing in new technology incrementally, slowly working up the SAE scales.

There is therefore a chance that we may become familiar with, and users of, Level 3 vehicles in the near future. A user of a Level 3 CAV (whom we might traditionally think of as the 'driver') is required to monitor the CAV at all times, and be ready to override or intervene in its driving, once more assuming the 'traditional' driver role. The Law Commission have suggested that an individual monitoring a CAV be referred to as a "User in Charge"[1], and that phrase has been adopted throughout this book.

As we will explore, the combined responsibilities of a User in Charge and a CAV pose practical and legal problems, particularly where Level 3 vehicles are concerned. The concerns surrounding Level 3 CAVs have even caused some to call for their prohibition on public roads, or for their use to be restricted to certain environments[2].

Level 3 CAVs are highly unlikely to be governed by the Automated and Electric Vehicles Act 2018 ("AEVA"): Section 1 of the AEVA refers to vehicles which are "*...designed or adapted to be capable, in at least some circumstances or situations, of safely driving themselves*". By their very definition, Level 3 CAVs can never guarantee that they can safely drive themselves: they require constant monitoring by a User in Charge, and can request that the user takes over the driving at any given time. This would appear to bring them outside of the AEVA's definition and coverage. As a result, it seems inevitable that our current legal framework would need to be adapted to accommodate road traffic collisions involving Level 3 CAVs. This is not without its challenges, which we will explore in this chapter. We will also explore the issues which are likely to be at the heart of any future disputes, and the ways in which our legal system may need to develop to resolve these problems.

At the time of writing (in September 2019), there are two significant unknowns which will have a substantial bearing on the role and responsibilities of a User in Charge, as well as the outcome of any negligence claims concerning CAV collisions.

---

[1] The Law Commission's Consultation Paper No. 240 "Automated Vehicles: A Joint Preliminary Discussion Paper", dated 8 November 2018, particularly at paragraph 1.42: https://s3-eu-west-2.amazonaws.com/lawcom-prod-storage-11jsxou24uy7q/uploads/2018/11/6.5066_LC_AV-Consultation-Paper-5-November_061118_WEB-1.pdf

[2] See, for example, the Law Commission's "Analysis of Responses to Law Commission Consultation Paper No 240", dated 19 June 2019, at paragraphs 3.131 – 3.136: https://s3-eu-west-2.amazonaws.com/lawcom-prod-storage-11jsxou24uy7q/uploads/2019/06/Automated-Vehicles-Analysis-of-Responses.pdf

The first unknown is whether and when a User in Charge will be required to intervene in a CAV's driving. Will he or she only need to do so in response to a request from the CAV itself (for example, where it has encountered a road without connectivity, or which is otherwise beyond its capabilities)? Or will Users in Charge be expected to voluntarily intervene in, or override, the CAV's driving (for example, where they believe that a CAV is driving inappropriately or dangerously)? This is explored in more detail later in this chapter, under the heading 'The Duty of Care and Standard of Care'.

The second (and related) unknown is the extent to which Users in Charge can reasonably be expected to monitor a Level 3 CAV. It will be challenging to maintain focus and avoid distraction whilst a Level 3 CAV has responsibility for the task of driving: one can imagine how easy it might be to let one's thoughts drift off, or to be tempted to quickly answer a call or text message.

The courts are unlikely to condone such wilful neglect of one's duty to monitor a Level 3 CAV. Might they be more forgiving of a User in Charge who is still watching the road, but has lost their focus? This seems unlikely: if Level 3 CAVs are to function safely on public roads, it seems that the courts would have no choice but to hold users to high standards. Accepting that Users in Charge will reasonably lose their focus and attention would, in effect, be an acceptance that Level 3 CAVs are not safe for general use. In this regard, manufacturers will also need to play their part by assisting Users in Charge; for example, they may need to install regular in-car alerts or other attention-grabbing devices.

In light of the unknowns surrounding Level 3 CAVs, the remainder of this chapter has necessarily proceeded on the basis of three assumptions:

1. Level 3 CAVs will be manufactured and permitted to be driven on public roads;

2. Users in Charge of Level 3 CAVs will be expected to maintain their focus and concentration on the CAV's driving at all times, as if he or she were the driver themselves;

3. Users in Charge of Level 3 CAVs will be expected to intervene in a CAV's decisions and driving, both on request and voluntarily.

BRINGING PROCEEDINGS AGAINST A DRIVER IN A NON-AEVA CASE

Today, those who have the misfortune to be involved in a road traffic accident are likely to be aware of their right to bring proceedings against the 'at fault' driver (or of the rights of others to bring proceedings against them, in the event that they were the driver at fault). They may or may not know that this involves suing the driver for negligence, or that their insurance company may be doing this on their behalf.

What is likely to be well known to all readers of this book, but is briefly set out here for completeness, is that a party bringing negligence proceedings would need to prove:

- That the driver owed them a duty of care (although this is rarely a contentious point, as the common law has long accepted that drivers owe a duty of care to other road users);

- That this duty of care was breached; i.e. the driver's actions fell below those expected of a reasonable driver;

- That this breach of duty caused injury, loss or damage.

The application of the principles of negligence to road traffic accidents is a well-trodden path. As a result, the majority of cases can be resolved on the facts of the case alone, and without explicit reference to (or debate on) the legal principles of negligence. To date, these familiar approaches have, of course, been based on conventional (Level 0) vehicles. As we will now discuss, applying them to CAV technology is not without its problems.

The Duty of Care and Standard of Care

Users in Charge of CAVs will still owe other road users a duty of care: the introduction of advanced technology is unlikely to justify a departure from the status quo (and, if anything, its new and inherently dangerous nature is more likely to justify an enhanced duty of care).

The waters begin to get muddied when we turn to the *standard of care* that will be expected of Users in Charge, i.e. a court's assessment of whether their conduct has fallen short of that expected of a reasonable User in Charge.

CAV technology will – for quite some time – be in a state of continuous development, and this is likely to induce an element of user unfamiliarity or hesitancy. In that context, how will judges determine what is 'reasonable'? Issues to be determined by the courts include:

- At what point would a court declare that a reasonable user should have intervened in the driving of a Level 3 CAV?

- Would the absence of a warning or request from the CAV be sufficient to exonerate their blame?

- Would failing to read a CAV instruction manual from cover to cover be deemed unreasonable?

- In what circumstances and environments would it be unreasonable to deploy an automatic parking feature?

Overriding a Level 3 CAV: what is reasonable?

The expectations surrounding voluntary intervention are particularly tricky to assess, and can all too easily be judged with the benefit of hindsight. Imagine if you were a User in Charge monitoring your new Level 3 CAV. You are travelling at 50 miles per hour in the middle lane of a dual carriageway. The CAV has control of all driving functions, and the in-built monitoring system shows that it will soon transition to the left-hand lane, ready to exit the dual carriageway at the next slip road.

You notice a small red car in the left-hand lane, marginally ahead of your CAV. Your CAV turns on its indicators, and begins to drift toward the left-hand lane. Your gut reaction tells you that this is a little too early, and that the CAV should have left more space between you and the red car. You reassure yourself that your CAV – with its sensors, lasers, cameras and high-tech GPS – will have calculated the speed of both vehicles, as well as the distance between them, and would not have proceeded if it were unsafe to do so. You also recall being told by the seller that these vehicles can communicate with each other using Vehicle to Vehicle ('V2V') technology: perhaps the car in front is also a CAV and the two vehicles have synchronised their movements?

Unfortunately your CAV moves to the left-hand lane but does not reduce its speed, and the gap between the vehicles begins to close. You reach for the mechanism which will enable you to take control of the vehicle, but it is too late. You collide with the red car, causing damage to both vehicles, and causing your passengers to sustain minor injuries. Should you be criticised for failing to intervene at the point where your gut told you something was wrong?

In a parallel world, at the very point that your CAV began to drift towards the left-hand lane, you decided to override the system and take back control. A few seconds later you cancelled the left indicator and returned to the middle lane. However, the blue car behind you saw you leaving the lane, and decided to accelerate to make progress. It could not reduce its speed in time, and crashed into the rear of your CAV.

Was your last-minute change of direction unreasonable? Should you have simply trusted the CAV to do its job?

In addition to voluntary interventions, a CAV-initiated handover could cause difficulties, for example, where a CAV gave very little warning of an imminent handover, or if the User in Charge had lost their focus, and was not sufficiently prepared to take action to avoid a collision.

In these scenarios, a court's assessment of the user's actions (whether they fell below the standard of a reasonable User in Charge), will be highly fact specific. Key factors will include:

- the presence or absence of a warning or request from the CAV;

- the amount of time or notice given by the CAV;

- reaction times;

- the ease or difficulty of activating the 'override' function in a particular CAV;

- the User in Charge's level of concentration on the road;

- the User in Charge's perception of the risk, and when he or she first became aware of the risk;

- any factors against human intervention, for example dense fog, which might make the sensors of a CAV more reliable than the human eye.

As with the case law surrounding conventional vehicles, it will take some time for the courts to establish the norms we expect of Users in Charge in these complicated scenarios. This will necessarily result in a period of uncertainty for drivers, those seeking compensation, and insurance companies.

Should Users in Charge be expected (and permitted) to override Level 3 CAVs?

In light of the uncertainties explored above, should voluntary interventions be prohibited? Whilst it seems unlikely that CAVs will be developed in a way that physically restricts such an intervention, it would be possible for non-authorised interventions to be actionable in negligence proceedings.

If a major benefit of CAV technology is the elimination of human error from driving[3], it is illogical to place an obligation on humans to second guess the decisions of their CAV. The chances are that humans will alter a CAV's decisions out of mistrust or an abundance of caution, and will do so with significantly less information than the CAV (bearing in mind its sensors and its ability to connect to the surrounding environment). Not only will this undermine the benefits of CAVs, it could in fact *create* a dangerous situation (as with our blue car example above).

Furthermore, the quality of human intervention is likely to be questionable: as humans begin to regularly use and rely upon CAVs, their driving experience, reactions, and perceptive abilities will decline. This will be particularly true of new drivers, who will not have had the opportunity to cement their repertoire of new skills. Thus, when drivers are called upon to intervene or take control of the driving, typically in 'difficult' or emergency scenarios, their skills will be less fine-tuned, yet more important than ever.

Despite this, the argument that humans should be prohibited from playing an active role in minimising the risks of CAVs will strike many as unpalatable. This is particularly so when one acknowledges that

---

[3] To fully understand the significance of this, see the House of Lords Science and Technology Select Committee's Paper No. 115 'Connected and Autonomous Vehicles: The future?', dated 15 March 2017. The evidence given to the committee was that human error was a causal factor in 90-95% of road traffic accidents (see paragraph 81): https://publications.parliament.uk/pa/ld201617/ldselect/ldsctech/115/115.pdf

initial CAV technology will be far from perfect, and that errors and accidents will almost certainly occur.

The AI behind CAVs will necessarily be far more complex than the algorithms used in other forms of automated transport (which may be as simple as 'green light = go'; 'red light = come to a stop and remain stationary'). CAVs must instead interpret multiple sources of data, and make a considered decision as to the best way to proceed. Sticking with the rudimentary example of traffic lights, a CAV must, as a minimum, do all of the following: use its camera to assess the traffic lights; ascertain its location on a GPS; and scope the area for any vehicles, pedestrians or objects which may cause a collision (using a combination of its camera and other sensors, such as lidar, radar and/or ultrasound). The CAV must then proceed in a manner and speed that is appropriate for the surroundings, including the weather. Such a complex composition of facts, all liable to change within seconds, cannot be pre-programmed. The CAV must use its own artificial intelligence to make split-second decisions, and to continually learn from them.

Whilst the testing and programming of CAVs will undoubtedly be rigorous, the inability to pre-programme all situations, or to control the AI's learning, makes it inevitable that the debut of CAVs will necessitate a period of reflection and learning. CAVs may initially make poor decisions, or fail to recognise objects or dangers, which will, in turn, endanger its occupants.

This 'trial period' will be most prominent in Level 3 vehicles, assuming that they will be permitted on our roads. This, in itself, may be a compelling reason to require Users in Charge to intervene in a CAV's driving where they deem it necessary. Alternatively, the dangers associated with a 'trial period', combined with the uncertainty and poor quality of human interventions, may be a compelling argument for a prohibition on Level 3 CAVs altogether.

The Law Commission's Preliminary Consultation Paper[4] proposed that where the

> "...automated driving system [is] appropriately engaged, the user-in-charge would not be in control of the vehicle. They would not be responsible for any problems arising from how the vehicle is driven."

It would appear to follow from this that there would be no requirement on a User in Charge to *voluntarily* intervene in a CAV's driving. The simplicity of this approach is attractive, as it enables us to bypass many of the problems set out above, and to allocate accidents as either 'user-caused' (and dealt with through typical negligence proceedings) or 'vehicle-caused' (dealt with through the channels of product liability, which are discussed further in Chapter 9).

However, whilst this approach would work well for Level 4 and Level 5 CAVs (where there is a clear dividing line between the responsibilities of the CAV and the User in Charge) it seems inappropriate in the context of Level 3 CAVs: if a User in Charge is required to monitor the CAV at all times, surely their monitoring (or lack thereof) ought to be subject to legal scrutiny? A lack of responsibility and accountability could lead to a sense of complacency, creating (or enhancing) the risk of Users in Charge failing to give a CAV's driving their undivided attention.

Causation and Contributory Negligence before the AEVA 2018

Causation

The difficulties do not end there for claimants bringing negligence proceedings: they will also need to establish both factual and legal causation. This book is not the appropriate forum to explore the breadth and depth of case law on tortious causation; our endeavour is to provide examples where CAVs may throw up complications.

---

4 Footnote 1, at paragraph 3.47

Returning to our earlier example of your new CAV and the red car: if you *had* decided to intervene in the CAV's decision, would it have made a difference? Allowing for your personal reaction time, the time taken to switch your CAV to its 'manual' mode, and the time it took you to return your CAV to the middle lane, it may nonetheless have been too late. The accident may still have occurred, though it may have caused less damage.

These questions, aimed at establishing factual causation, will be highly fact-specific. Their resolution is also likely to require a myriad of complex evidence, including:

- information on reaction times;

- an explanation of the speed and ease with which a particular model of CAV can switch from 'automated' mode to 'manual' mode;

- information from the system as to whether or not a warning or request for intervention was issued; and

- the various forms of 'live' information and data captured by the CAV.

The resolution of road traffic accidents is likely to become significantly more technical and expensive as a result.

Contributory Negligence

Allegations of contributory negligence are likely to present additional complexities. Let us take a new example: you are the User in Charge of your new Level 3 CAV which is driving itself along a country lane. On a stretch of road you see a cyclist. There is sufficient room to overtake the cyclist, and the other side of the road is clear. As your CAV begins to overtake of its own volition, the cyclist veers out into your path. You have been monitoring the CAV's driving, and your instincts cause you

LIFE BEFORE THE AUTOMATED AND ELECTRIC VEHICLES ACT 2018 • 25

to grab the steering wheel and direct the CAV away from the cyclist. However, a vehicle approaching from the opposite direction is driving too fast. It emerges round a bend and crashes into the front of your CAV.

Who is to blame here: the cyclist for veering out? Your CAV for failing to pre-empt this situation, and/or failing to respond in the heat of the moment? You, for taking your car onto the wrong side of the road, and possibly undermining the CAV's response? The speeding driver approaching from the opposite direction? Or a mixture of all of the above and, if so, in what proportion? A judge seeking to resolve this situation will be faced with conundrums of principle, factual untangling and analysis, and - in all likelihood - an abundance of technical data.

This multi-party confusion also brings uncertainty as to who should be sued, and for what. The above example involved four potential parties: you; the cyclist; the manufacturer of your CAV; and the oncoming driver. The roles of insurers, and the garages who installed/checked/amended your CAV's software, may also need to be considered[5]. It would therefore seem that claimants falling outside the scope of the AEVA regime could have a difficult, expensive, and protracted route to receiving compensation.

Another facet of CAV technology which may further complicate these matters is the possibility of human intervention taking place *outside* the CAV, for example, an employee remotely monitoring a delivery vehicle, ready to assume the controls of the CAV if required to do so.

The Government's latest code of practice on automated vehicle trialling (produced by the Centre for Connected and Autonomous Vehicles)[6] confirms that trials of Level 3 CAVs may be conducted remotely,

---

[5] Product liability actions against these parties are explored in more detail in Chapter 9

[6] The Centre for Connected and Autonomous Vehicles' 'Code of Practice: Automated vehicle trialling' dated February 2019: https://assets.publishing.service.gov.uk/government/uploads/system/uploads/attachment_data/file/776511/code-of-practice-automated-vehicle-trialling.pdf

provided that minimum safety requirements are met. Although it is not yet clear whether remotely monitored CAVs will be permitted for mainstream use on public roads, this could introduce yet another level of complexity to the causation analysis. Questions posed by this technology include:

- How quickly would one be expected to act when removed from the situation?

- How would the distance impact one's reaction times?

- Would commands take longer to register with the CAV given the physical distance between the CAV and the monitoring individual?

- What happens when signal or connectivity fails, and who would be responsible for any injury or damage as a result of such failures?

Whilst the Law Commission's analysis of responses to its preliminary consultation paper suggests there would be a separate regime for remotely operated CAVs[7], it is nonetheless likely that the courts would still need to confront these questions in the context of negligence proceedings.

The questions posed in this section are not exhaustive. We foresee a potential minefield of litigation surrounding the standards of conduct and the rules of causation on our roads, as our older legal principles adapt to new technology and societal behaviours.

---

7 The Law Commission's "Analysis of Responses to Law Commission Consultation Paper No 240" (footnote 2), particularly at paragraphs 1.7 and 3.11

Negligence Actions Post-AEVA?

Although this chapter has primarily been aimed at Level 3 CAVs (and any Level 4 or Level 5 CAVs not covered by the AEVA[8]), it should be noted that the problems we have discussed could – in the absence of additional changes to the law – apply to any CAV. This is for three main reasons.

Firstly, the AEVA does not exclude or prevent claimants from suing other parties in negligence, rather than relying on the AEVA. Whilst this is inherently unlikely in light of the advantages offered by the AEVA (explained in detail in Part Two of this book), it is theoretically possible.

Secondly, whilst the AEVA gives claimants the opportunity to bypass the thorny problems raised in this chapter, such problems may nonetheless be of ongoing concern to insurers. Where insurers seek to recover their outlay from others (for example, the User in Charge, or another driver), they will still be faced with the complex factual and legal problems examined in this chapter.

Thirdly, some CAV-related claims may fall outside of the AEVA regime, for reasons explored in Chapter 5 ("Exclusions and restrictions in the AEVA 2018"). Absent any further legislation on this topic, such claims are most likely to be dealt with as negligence actions, requiring claimants to face the issues highlighted in this chapter.

Thus the factual and practical problems associated with CAV-related negligence claims will not disappear overnight, even with the assistance of the AEVA.

Actions against manufacturers and software installers

The bulk of the examples in this chapter have assumed that, at least to some extent, the 'User in Charge' of a Level 3 CAV was at fault. But

---

8 On which see Chapters 2 and 5

what if the CAV technology had a part to play, or was in fact the entire cause of the mishap? Such situations might call for a product liability claim to be made against a manufacturer, seller, or other supplier. We discuss these claims in Chapter 9.

HOW FUTURE LAWS MIGHT DEAL WITH OUR CASE

Could – and should – Parliament legislate to explicitly identify the standards to which Users in Charge of Level 3 CAVs will be held? It could certainly do so, but like any area of codification the legislation could never provide for every factual scenario. This seems especially so in a world of rapid developments and self-learning AI, where legislation is likely to become outdated swiftly, and would require constant monitoring and revision.

Even if one sets these difficulties aside, legislation could never provide for the practical uncertainties faced by those bringing negligence claims; for example, the complex factual analysis, the sheer volumes of information, and the number of parties that may need to be involved in a legal claim.

On any view, it is incontrovertible that the courts will be required to adjudicate on this subject matter, and that our legal system will face fresh challenges of a kind which may call into question the very way we litigate.

SUMMARY OF POINTS ON DRIVING CAVS BEFORE THE AEVA

Level 3 CAVs are yet to be manufactured for widespread use on our public roads, but there is a distinct possibility that they may appear within the next few years.
Accidents in Level 1-3 CAVs will be governed by the principles of negligence
It is unclear whether Users in Charge will be expected (or permitted) to voluntarily intervene in a CAV's driving
CAV handovers and interventions may lead to accidents. It will be for the courts to assess whether the actions of a User in Charge were 'reasonable'
Where a CAV has control of the driving functions, a User in Charge could nonetheless be found negligent for failing to monitor the driving
Questions of causation and contributory negligence will require detailed factual analysis, which is likely to necessitate the provision of complex evidence
The AEVA cannot resolve all of the problems associated with CAV-related negligence claims, and the courts will be required to be determine these difficult issues

PART TWO

THE AUTOMATED AND ELECTRIC VEHICLES ACT 2018

ALEX GLASSBROOK

CHAPTER TWO
OVERVIEW OF THE
AEVA 2018

INTRODUCTION

When the first edition of this book ("The Law of Driverless Cars: An Introduction") was published, in February 2017, there was no Act on the statute book in relation to autonomous vehicles. The Automated and Electric Vehicles Bill became the Act of the same name when it received royal assent on 19 July 2018. As at the time of writing, however, the Part of the Act dealing with automated vehicles has not yet been brought into force.

OBJECTIVES OF THIS CHAPTER

The objectives of this chapter are to:

- familiarise the reader with Part I of the Automated and Electric Vehicles Act 2018;

- explain which areas of activity Part I of the Act legislates, and which it does not;

- explore the sections in Part I of the AEVA 2018 which have raised queries, to highlight questions which will be dealt with in later chapters.

OVERVIEW OF PART I OF THE AEVA 2018

The AEVA 2018 – specifically Part I of the Act, entitled "Automated Vehicles: Liability of Insurers etc" – does not deal with all of the legal topics raised by CAVs. Far from it: of the 12 areas which the British

Government's Department of Transport had identified in its 2015 paper, reviewing "existing UK regulations and legislation to examine their compatibility with automated vehicle technologies"[1], the AEVA 2018 deals with only one: insurance (and particularly the "liability of insurers").

That approach was deliberate – to take "a step-by-step approach ... regulating in waves of reform" so as to be able "to learn important lessons from real life experiences of driving of increasingly automated vehicles"[2] – and dealt with what government perceived as the first concern relating to advanced automated vehicles: that those injured by accidents caused by CAVs while operating in automated driving mode would not be effectively compensated.

That was because the existing system of compulsory, third party road traffic accident insurance, against injury and damage caused by human drivers of conventional vehicles, was ill-fitted to automated vehicles operated by their own, artificially intelligent systems.

In relation to conventional vehicles, the injured claimant is put to the task of proving liability against the culpable defendant driver. If the claimant succeeds, the insured defendant's motor insurer pays the judgment (either directly, if sued under the rights against insurers regu-

---

1 UK Department of Transport, Centre for Connected and Autonomous Vehicles (CCAV), "Pathway to Driverless Cars: a detailed review of regulations for automated vehicle technologies", February 2015, at https://assets.publishing.service.gov.uk/government/uploads/system/uploads/attachment_data/file/401565/pathway-driverless-cars-main.pdf

2 CCAV, "Pathway to Driverless Cars: Proposals to support advanced driver assistance systems and automated vehicle technologies", July 2016 page 8, at https://assets.publishing.service.gov.uk/government/uploads/system/uploads/attachment_data/file/536365/driverless-cars-proposals-for-adas-and_avts.pdf

The Government decision, following the consultation, maintained that approach: page 9 of the Government response to the consultation, January 2017, at https://assets.publishing.service.gov.uk/government/uploads/system/uploads/attachment_data/file/581577/pathway-to-driverless-cars-consultation-response.pdf

lations[3], or indirectly, by the section of the Road Traffic Act requiring the motor insurer to satisfy the judgment against its insured[4]).

In a future claim where damage is caused by a fully self-driving vehicle, a system based upon the law of product liability would introduce difficulties. For example:

- At the first stage, the issue of liability would turn not upon the established law of negligence and the standard of the reasonably prudent driver[5], but instead largely upon the law relating to product liability, which concerns the level of safety in products which society is entitled to expect. The law of product liability is not, in its current form, easily adaptable to injury claims involving machine intelligence and vehicles which operate through a combination of sophisticated hardware, software and regular software updates. For example, product liability remedies do not (as the law stands) apply to a fault in software updated over a wireless connection, without any physical medium such as a disc[6]. That is clearly inapplicable to modern mobile technology, which will be essential to CAVs.

- At the second stage (insurance), the law of compulsory, third party motor insurance which developed to maturity during the

---

3   At the date of writing, the European Community (Rights against Insurers) Regulations 2002 (SI 2002/3061).

4   Section 151 of the Road Traffic Act 1988.

5   "He must drive in as good a manner as a driver of skill, experience and care, who is sound in wind and limb, who makes no errors of judgment, has good eyesight and hearing, and is free from any infirmity" (*Nettleship v Weston* [1971] 2 QB 691 at 699-700). "Sound in wind and limb" is an equestrian phrase, meaning in good health ("wind" is breathing). Lord Denning MR explained the high standard of care applicable to the driver of a car: "The reason is so that a person injured by a motor car should not be left to bear the loss on his own, but should be compensated out of the insurance fund. The fund is better able to bear it than he can. But the injured person is only able to recover if the driver is liable in law. So the judges see to it that he is liable, unless he can prove care and skill of a high standard". So liability and insurance questions are intertwined.

6   See the discussion in Chapter 9: "Product liability claims".

twentieth century presumes an identifiable, human driver. In that system, proceedings must be brought either against that human driver or, in cases where the insurer does not dispute the insurance coverage, directly against that driver's insurer[7]. By contrast, the essential feature of a Level 5 CAV would be its ability to operate without a human driver. The obligation to carry third-party motor insurance applied to people, not to systems nor to their manufacturers. So establishing fault solely on the basis of product liability, and compulsory insurance of the risks of product liability, would not be an easy matter either for claimant or insurer (see Chapter 9 – "Product Liability Claims").

That is not an exhaustive list of the unsuitabilities of the current civil liability regime to CAVs, but those are prominent reasons. The existing system rested upon the law of negligence, and would not fit CAVs.

The AEVA 2018 tackled the problem by cutting through the two stages.

The AEVA 2018 used an existing idea (direct right of action against an insurer, which the 2002 regulations allow) and extended it straight to its natural destination: the motor insurer. So the AEVA 2018 creates a direct, single-step liability against an insurer. As section 2(1) provides:

"*Where—*

(a) an accident is caused by an automated vehicle when driving itself on a road or other public place in Great Britain,

(b) the vehicle is insured at the time of the accident, and

(c) an insured person or any other person suffers damage as a result of the accident,

---

7   European Community (Rights against Insurers) Regulations 2002 (SI 2002/3061).

the insurer is liable for that damage."

The flipside of the Act's extension of insurers' liabilities is the extension of the insurers' *rights* in relation to CAV claims, allowing insurers to reduce or even extinguish their AEVA 2018 liabilities:

- by reference to the contributory negligence of the injured party (section 3),

- by contribution from any other liable party (section 5),

- both of which echo insurers' existing rights, and:

- by reference to uninsured acts or omissions in updating software to the automated vehicle (section 4).

A further balance of the insurers' liabilities under the Act came in the government's interpretation (as per section 8 of the Act, discussed below) that the Act would not apply to Level 3 vehicles (the "murky middle" of CAV liability, as the Law Commission puts it, where the robot system carries out the driving for most of the time and in circumstances familiar to it, but the human driver must remain available to intervene when appropriate, to bring the vehicle to a safe state[8]). The effect of that interpretation is controversial, as it would remove the strong remedy of the AEVA 2018 from any person injured in a RTA involving a Level 3 vehicles, and leave them seeking other legal remedies (see the discussion of that aspect of the Act in Chapter 5 - "Exclusions and restrictions in the AEVA 2018").

At the time of writing, the provisions of the AEVA 2018 governing automated vehicles have yet to come into force. They await the decision of the Secretary of State to activate the Act, including by stating to which new vehicles the Act applies (see below).

---

8   See summary of the SAE levels of automation in the Introduction, above.

Whether the AEVA 2018, as enacted, will, in practical terms, achieve the aim of extending compulsory insurance cover to those foreseeably affected by accidents attributable to CAVs is not yet tested. At the time of writing, the AEVA 2018 has not been tested in court: The vehicles have not yet appeared on the roads and the Act has not been brought into force.

The reason for that uncertainty is the very slight extent to which the AEVA 2018 describes the changes which connected and automated vehicles will bring to liability questions.

The AEVA 2018 and Liability

The 2018 Act makes clear that it provides for liability in tort. section 6(4) ("Application of enactments") provides that:

> *"Liability under section 2 is treated as liability in tort or, in Scotland, delict for the purposes of any enactment conferring jurisdiction on a court with respect to any matter."*

The AEVA 2018 aims to settle the insurance question, but it leaves the underlying liability questions, which determine whether or not any compensation is due in the first place, largely to the courts, with some clues as to how to decide those liability questions.

Those questions are mainly generated by the factual difficulties of deciding causes in road traffic accidents. Those difficulties exist in abundance in current RTA claims: accident reconstruction is increasingly required, especially because of the sophistication of modern vehicle design, which makes interpretation of damage a task in which the judge often needs expert help. The increase in sources of digital information adds further complexity.

While recognising its function in cases requiring scientific opinion, the courts have resisted the introduction of expert accident reconstruction evidence to a broad swathe of road traffic accident cases – chiefly to reduce legal costs. But the AEVA 2018 raises the issue of causation of accidents, in a multi-vehicle, multi-technology setting. So trials and evidence – both its nature (oral and documentary, lay and expert, civilian and police), volume and the ways in which it is presented to the courts (paper and digital) – will change.

The AEVA 2018 adopts the policy of leaving factual matters to the courts. It nevertheless treads upon the ground that it purports to leave open to the judges. In legislating for the extension of compulsory insurance to accidents caused by CAVs, the AEVA 2018 has been unable to avoid speaking of causation, of contributory negligence and of contribution between tortfeasors (even creating its own, particular remedy of contribution for the insurer in section 5, and disapplying for that purpose the Civil Liability (Contribution) Act 1978[9]). section 8(3) of the Act provides two particular principles of causation, in relation to the statutory right of direct action which it provides under section 2 (discussed below).

So, to describe the Part I of the AEVA 2018 as being confined to insurance questions and not dealing *at all* with liability, would not be precisely accurate. The AEVA 2018 provides for a tortious liability (see section 6(4), above). It even starts to set the terms of that liability (see section 2). But it sets the terms relatively vaguely.

It might be more accurate to say that the AEVA 2018, in establishing a particular scheme of direct liability of insurers in relation to CAV accidents, sets the rules of that scheme, and leaves the running of the scheme to experience and the interpretation of its rules to the courts.

In particular, the question of how a judge is to understand and identify robotic reasoning, as distinct from human reasoning, in order to distinguish the two and to fix an appropriate share of fault where both sides have contributed, is factually and legally difficult. But, in the first place, whether or not the AEVA 2018 requires such a comparison of human and robotic reasoning is questionable. We will return to that issue.

---

9 AEVA 2018 section 6(5) ("Application of enactments")

Considering how the AEVA 2018 will direct the courts' approach to the underlying liability questions in CAV cases raises many interesting and difficult questions for the future.

We discuss those problems in more detail later in this chapter, and they will recur in different settings throughout the book.

SUMMARY OF THE PROVISIONS OF THE AEVA 2018 RELATING TO AUTOMATED VEHICLES

The AEVA 2018 describes itself simply as "An Act to make provision about automated vehicles and electric vehicles". Its provisions about automated vehicles appear in Part I of the Act, which is entitled "Automated Vehicles: Liability of Insurers etc".

At the time of writing, the Secretary of State for Transport has yet to make regulations to bring Part I of the AEVA 2018 into force.

There are eight sections in Part I of the Act. The list of titles of those sections from the Act is as follows, with our summary of the main features of each section. In summarising these provisions, we start to see some of the legal issues they will generate.

1. **Listing of automated vehicles by the Secretary of State:** the mandatory duty of the Secretary of State for Transport (SoST) to list all motor vehicles which are, in the SoST's opinion, designed or adapted to be capable, in at least some circumstances or situations, of safely driving themselves, and which may lawfully be used when driving themselves, in at least some circumstances or situations, on roads or other public places in Great Britain. Defines "automated vehicle" in Part I AEVA 2018 as meaning "a vehicle listed under this section": s.1(4)) (As discussed in Chapter 5 – "Exclusions and Restrictions in the AEVA 2018" – this definition appears to exclude Level 3 vehicles from the Act).

2. **Liability of insurers etc where accident caused by automated vehicle**: the keystone section of Part I of the AEVA 2018, providing for direct liability of the insurer of an AV for damage where an accident is caused by an insured AV when driving itself on a road or other public place in Great Britain (see Chapter 3 – "Causation in the AEVA 2018").

3. **Contributory Negligence etc**: section 3 preserves the reduction of damages for which the insurer is liable, by reference to any contributory fault of the injured party. Sets the standard by which the fault of an automated system will be assessed, in the balancing of contributory faults, as that which "would apply to a claim in respect of the accident brought by the injured party against a person other than the insurer or vehicle owner". The Law Commission and others have challenged the clarity of that language. See Chapter 4 – "Contributory negligence in the AEVA 2018".

4. **Accident resulting from unauthorised software alterations or failure to update software**: section 4 allows the insurer to avoid or limit its liability under section 2 by excluding or limiting cover in the policy for damage suffered by an insured person arising from an accident occurring as a direct result of prohibited software alterations made by the insured person, or a failure to install safety-critical software updates that the insured person knows, or ought reasonably to know, are safety-critical. See Chapter 5 – "Exclusions and Restrictions in the AEVA 2018".

5. **Right of insurer etc to claim against person responsible for accident**: section 5 allows the insurer to seek contribution for its AEVA 2018 liability to the injured person, where the amount of that liability has been settled, from any other person who is liable to the injured person in respect of the accident. See Chapter 6 – "The Insurer or Owner's Right to Claim against Another Responsible Person under the AEVA 2018".

6. **Application of Enactments**: section 6 provides for the application of other Acts. Of note are (1) the exclusion (by section 6(5) AEVA 2018) of the Civil Liability (Contribution) Act 1978, if the insurer or owner seeks contribution under section 5 of the AEVA 2018) and (2) the extension of the section 2 AEVA 2018 liability to a child born disabled (per section 1 of the Congenital Disabilities (Civil Liability) Act 1976 – "civil liability to a child born disabled") as a result of injury to a parent in an AV accident.

7. **Report by the Secretary of State on operation of this Part**: This sets a mandatory requirement upon SoST to lay before Parliament, no later than two years after the first publication of the list under section 1, a report assessing the impact and effectiveness of Part I AEVA 2018 and the extent to which that Part's provisions "ensure that appropriate insurance or other arrangements are made in respect of vehicles that are capable of safely driving themselves".

8. **Interpretation**: The interpretation section in relation to Part I. Of particular note is the addition to the definition of "automated vehicle" in section 1(4) ("a vehicle listed" by SoST) of the further refinements that "a vehicle is "driving itself" if it is operating in a mode in which it is not being controlled, and does not need to be monitored, by an individual". That addition is relevant to the question of the types of vehicle (eg. as defined by the SAE levels) to which AEVA 2018 was intended by Parliament to apply (see the discussion of implied limitations to the Act, in Chapter 5).

Limitation periods for actions under the AEVA 2018 are set by Part 3 of the AEVA ("Miscellaneous and General"), section 20(1), and in the Schedule to the Act (by amendment of the Limitation Act 1980, setting limitation periods for the section 2 action against an insurer of 3 years and for the contribution action under section 5 AEVA of 2 years).

The Schedule to the AEVA also makes clear that it is not open to the owner of an automated vehicle, as an alternative to insuring the AV, to make a deposit of a sum with the Accountant General of the Senior Courts (by the AEVA's amendment of the Road Traffic Act 1988 to that effect - see paragraphs 17 and 18 of the Schedule).

The AEVA 2018 also, in effect, restores compulsory third party motor insurance cover for an insured employer's liability to its employee, whose death or injury arises out of and in the course of their employment, where the vehicle in use on a road other public place in Great Britain (the "vehicle in question") is an automated vehicle (Paragraph 19(3) of the Schedule, which in the case of an AV removes the exclusion at section 145(4)(a) of the Road Traffic Act 1988).

Under Part 3 AEVA, the Secretary of State for Transport has a power by regulations to make provision that is consequential on any provision made by the Act (section 20(2): "minor and consequential amendments) and to bring the Act into force (21(2): "Commencement"). The power to make regulations to commence the Act is exercisable by statutory instrument (ibid) and may appoint different days for different purposes or different areas (21(4)(a)).

A statutory instrument containing regulations under section 20 AEVA ("Minor and consequential amendments"), any of which amend, repeal or revoke primary legislation (which includes legislation of devolved assemblies: (20(7)) may not be made unless a draft of the instrument has been laid before Parliament and approved by a resolution of each House (20(5)). A statutory instrument containing regulations under section 20 none of which amends primary legislation is subject to annulment in pursuance of a resolution of either House of Parliament (20(6)).

At the time of writing, the Secretary of State for Transport has yet to publish regulations in relation to the AEVA 2018. In particular, SoST has made no regulations to bring the provisions of Part I of the AEVA 2018 into force. Nor has SoST laid the first list of automated vehicles before Parliament.

WHAT THE AEVA 2018 DOES – AND DOES NOT – LEGISLATE

The AEVA 2018 (in Part I) legislates the following areas:

- The responsibility of the government (the Secretary of State for Transport) to publish, and to review every 2 years after publication, a list of automated vehicles (sections 1 and 7)

- The liability of an insurer (or, in those cases where the Road Traffic Act 1988 does not require compulsory insurance, the owner of the vehicle) to a direct claim against it by the victim of damage suffered in an accident caused by an insured, automated vehicle when driving itself on a road or other public place in Great Britain (section 2)

- The insurer's rights, in relation to such a claim against it (section 3: "contributory negligence etc"; section 4: "Accident resulting from unauthorised software alterations or failure to update software"; section 5: "Right of insurer etc to claim against person responsible for accident"; the definition of "driving itself" in the interpretation section, 8(1)(a), is relevant to the debate as to whether or not a vehicle which might be described as SAE level 3 would fall under the AEVA 2018).

The AEVA 2018 does NOT legislate the following areas:

- The liability of the insurance industry in relation to any accident involving an *un*insured CAV (an uninsured conventional vehicle would fall to be considered under the Motor Insurers' Bureau agreements relating to untraced or uninsured drivers. The MIB is not a statutory body. At the time of writing the MIB has not stated its position in relation to accidents involving an uninsured CAV).

- The standard of care applicable to a person in charge of a CAV, particularly at Levels 4 and 5 (who the Law Commission has proposed should be called "the user in charge" – see Chapter 1). Whether such users should be held to a higher standard of care than drivers of conventional vehicles is a topic upon which the Law Commission is consulting. The authors note that the answer to this question might alter with context (the role of the UIC might be, for example, employee, passenger or carer).

- The AEVA does not set the standards of safety we can expect of CAVs, nor the standards that manufacturers will be held to (see Chapter 9, "product liability claims").

- The AEVA does not provide for recovery against the insurer for the damaged CAV (see section 2(3)(a)), for which a claimant would need fully comprehensive insurance and/or another remedy (such as a product liability claim).

- Data security, privacy or confidentiality issues relating to CAVs (for example, arising from the use by CAVs of personal data such as location and financial data).

- Criminal offences relating to CAVs.

THE SECTIONS OF THE AEVA 2018 WHICH HAVE RAISED QUERIES

The following areas are legislated by the AEVA 2018, but their meaning has been questioned:

- The approach to causation in an AEVA 2018 claim (including how a claim involving several vehicles – some insured AV's giving rise to an AEVA 2018 claim for victims, but some not – would be adjudicated at trial);

- The approach to contributory negligence in an AEVA 2018 claim (including the standard of care to be applied to a human driver in an accident involving an automated vehicle);

- The approach to contribution under the AEVA 2018 from other tortfeasors liable to the victim in relation to the same accident (particularly given the departure of the right of contribution under the AEVA 2018 from the 1978 Act);

- How the exclusion of AEVA liability for breaches in relation to software might operate in practice; and

- Which SAE Levels of AV's (4 or 5, as stated by the government, or 3, 4 and 5) are covered by the AEVA 2018.

The debate is at too early a stage to present an exhaustive list of queries in relation to the AEVA 2018, and no answer can be certain at this early stage.

But it is important to start mapping the way. We embark on that journey in the following chapters.

SUMMARY OF POINTS ON AN OVERVIEW OF THE AEVA 2018

The AEVA 2018 is concerned mainly with insurance coverage, to reassure those injured by AV accidents of a means to compensation.
The AEVA 2018 seeks to provide that reassurance by providing a direct right of action against the insurer of an insured AV, where an AV caused or partly caused the accident.
But, to create that right, the AEVA 2018 cannot avoid dealing with some legal liability issues.
The Act treads lightly on legal liability questions. But even its light tread creates ripples in sensitive legal areas (causation, contributory negligence, coverage of uninsured CAVs etc).
How those ripples will develop is unclear, because the practical effects of automated vehicle technology are unclear. Some of the ripples from the AEVA 2018 might turn into legal waves.
We deal with those questions in the rest of the book.

CHAPTER THREE
CAUSATION IN THE AEVA 2018

INTRODUCTION

As discussed in the previous chapter, for a victim of an accident involving an automated vehicle, the key to unlocking the strong remedy in the 2018 Act (once the Act has been shown to apply to the vehicle) is likely to be proving that "an accident [was] caused by an automated vehicle when driving itself on a road or other public place" in Great Britain (section 2(1) AEVA 2018).

So causation of the "accident" in question is key to proving liability under the AEVA 2018.

This chapter deals with that question: causation of the accident. The other question of causation under section 2(1) – the requirement that "an insured person or any other person suffers damage as a result of the accident" (2(1)(c)) is more familiar and unlikely to generate as much controversy as the question of whether or not the accident was caused by the autonomous car. So the question of causation of damage is not the subject of this chapter. "Damage" is defined as including death or personal injury and property damage, with the latter qualified by certain categories and value (see sections 2(3) and (4)).

So, to return to the question of causation of the accident:

"Caused by" includes "partly caused by" (section 8(3)(b)) and "an accident" includes several accidents, provided they are "causally related" (section 8(3)(a)).

Those two refinements of causation in section 8 are familiar, common-sense principles. They appeal to what H.L.A Hart and Tony Honoré described 60 years ago (in the language of the time) as "the plain man's notions of causation (and not the philosopher's or the scientist's)"[1].

Transport affects everyone. But artificial intelligence is also the territory of the scientist and of the philosopher[2] (perhaps even of the seer, as Level 4 CAV technology has yet to arrive). Automated vehicles give rise to the most difficult questions of technical mechanics and of moral responsibility.

So it is unsurprising that the issue of causation under the AEVA 2018 – and how it is to be defined - has already generated a great deal of discussion[3].

For the reasons set out above (the apparently limitless scope of philosophical debate as to the legal effects of AI) and below (an analysis of the AEVA 2018 on its own terms), the writers suggest that the AEVA 2018 seeks to narrow the scope of causation questions. In other words, it seeks to limit the number of questions in an AEVA claim to a practical and manageable quantity. Whether or not it will achieve that aim remains to be seen.

In the first place, the AEVA 2018 asks a particular factual question, by asking whether the automated vehicle caused (or partly caused) "the

---

1 *Causation in the Law*, 1st edition, 1959, Oxford University Press, Part 1: "The Analysis of Causal Concepts"

2 See, for example, "Life 3.0" (Penguin, 2018) by the physicist Professor Max Tegmark and "Hello World: How to be Human in the Age of the Machine" (Bloomsbury, 2018) by the mathematician Dr Hannah Fry.

3 See the Law Commissions' (of England and Wales and of Scotland) preliminary consultation paper on Automated Vehicles (LC CP 240, SLC DP 166, 8 November 2018, pp108-110: https://s3-eu-west-2.amazonaws.com/lawcom-prod-storage-11jsxou24uy7q/uploads/2018/11/6.5066_LC_AV-Consultation-Paper-5-November_061118_WEB-1.pdf) and the summary of responses to the Law Commissions' consultation (https://s3-eu-west-2.amazonaws.com/lawcom-prod-storage-11jsxou24uy7q/uploads/2019/06/Summary-of-Automated-Vehicles-Analysis-of-Responses.pdf)

accident(s)". It tries to avoid the deeper currents of philosophical debate[4].

In the second place, the 2018 Act separates that threshold causation question ("did the automated vehicle cause the accident?") from later questions as to contributory negligence and (after settlement of the amount of the primary AEVA 2018 claim) the insurer or owner's right to seek contribution to its liability from others. Those later questions involve some attribution of fault to others. It is arguable that the Act introduces questions of culpability at those later stages. Those later questions are not discussed in this chapter but in the subsequent chapters on those topics.

OBJECTIVES OF THIS CHAPTER

This is a "Practical Guide" to the law as it stands, so this chapter concerns itself mainly with how the AEVA 2018 deals with causation.

At the time of writing, Part I of the AEVA 2018 has not been brought into force and no regulations have been published under that Part of the Act. There is no case law applying the AEVA 2018. So what follows is unavoidably speculative.

Subject to those limitations, in this chapter, we aim to:

- Familiarise the reader with the provisions of the AEVA 2018 relating to causation

- Discuss the boundaries of the causation issue under the 2018 Act

---

[4] If it does seek causal simplicity in its own field, the 2018 Act has a different approach to Hart and Honoré, who sought "some more complex principle or set of principles" to "guide, though not dictate" the use of causal language in the law as a whole. "The pains of unearthing these, though considerable, seldom go unrewarded" (Causation in the Law, 2nd edition, 1985, p.3).

- Raise questions as to how the courts might approach the key question of whether an AV "caused the accident" under the AEVA 2018.

THE PROVISIONS OF AEVA 2018 RELATING TO CAUSATION

As noted in the last chapter, the AEVA 2018 does not set out a complete scheme of causation; it leaves causative questions largely to the courts.

However, it is not accurate to say that it sets no framework for the causation issue. The language of the Act defines the core causation question in a particular way, by requiring proof that the injurious accident(s) were caused by an insured, "automated vehicle when driving itself on a road or other public place" in Great Britain (2(1)(a)).

The provisions of the Act defining that core causation issue are as follows:

Section 2 ("liability of insurers etc where accident caused by automated vehicle"). In particular, section 2(1) provides that:

"Where—

(a) an accident is caused by an automated vehicle when driving itself on a road or other public place in Great Britain,

(b) the vehicle is insured at the time of the accident, and

(c) an insured person or any other person suffers damage as a result of the accident,

the insurer is liable for that damage."

The key phrase is in section 2(1)(a): "an accident is caused by an automated vehicle".

"Caused by" includes "partly caused by" (**section 8(3)(b)**) and "an accident" includes several causally-related accidents (**8(3)(a)**).

As discussed later, in Chapter 5, the Act also allows for exceptions from and restrictions to the statutory liability. Those exceptions or restrictions might be described in causal language (eg. the accident was caused by the User in Charge's entire negligence in allowing the vehicle to begin driving itself when it was not appropriate to do so (Subsection (2) of section 3, "Contributory negligence etc", in which ss.3(2) excludes liability) or by the insured person making software alterations prohibited by the insurance policy, or failing to update software (s.4)). But only section 3(2) completely bars liability on a "contributory negligence etc" grounds, and the limitations under section 4 have different effects according to the identity of the AEVA claimant. So the exclusions and restrictions do not have a uniform causative effect, and might be better explored on their own terms rather than under the heading "causation" (see the subsequent chapters on those topics).

For present purposes, therefore, the writers approach the main causation question under the Act (causation of the accident by an AV when driving itself on a British road) separately from contributory negligence and those statutory exceptions and limitations.

THE BOUNDARIES OF THE "CAUSATION OF THE ACCIDENT" QUESTION UNDER THE AEVA 2018

In all cases, it is the "accident" (or accidents) which the automated vehicle must have caused or partly caused.

So the main causation question under the AEVA 2018 is "Did the automated vehicle cause, or partly cause, the accident(s)?".

Although this point is presently unregulated under the AEVA 2018 and is untested by the courts, it appears arguable that the Act itself does not require fault in the sense of blameworthiness or culpability. It merely requires that the AV caused or partly caused "the accident(s)". If that is correct, then questions of the moral culpability (and the difficult, underlying philosophical debate) might be irrelevant, for the purpose of assessing the insurer or owner's liability under section 2 AEVA 2018.

"Accident" is not defined by the AEVA 2018, nor by the other statutory source of compulsory motor insurance, the Road Traffic Act 1988. The Oxford English Dictionary defines "accident" in the broad, non-legal sense as "1. An unfortunate incident that happens unexpectedly and unintentionally. 2. something that happens by chance or without apparent cause" and (in philosophy) "3. a property of a thing which is not essential to its nature"[5].

The definition of "accident" in any particular case might depend upon the definition (if any) given to that term in the terms of the insurance policy insuring the particular automated vehicle. Where the definition of "accident" or "accidental" is contentious in a claim, where an insurance policy is engaged, the courts' approach has been to look to the definition in the terms of the insurance policy[6].

We cannot be prescriptive as to how a court would construe the word "accident" in any particular AEVA case. That would depend upon the particular facts of any case (probably including the terms of the particular insurance policy). As yet unwritten regulations, made under the AEVA 2018, might apply.

But this point might arise: the possibility that the courts will not construe "accident" as requiring fault in the sense of blameworthy or culpable behaviour on the part of the automated vehicle, when asking

---

5 From the 10th edition of the OED (1999).

6 See, for example, the definition of the different phrase, "accidental bodily injury", as construed by the Court of Appeal in a security contractor assault case, in a public liability (not a compulsory motor) insurance policy, in *Hawley v Luminar Leisure plc* [2006] EWCA Civ 18; [2006] Lloyd's Rep IR 307, §§94 to 119.

whether or not section 2 applies to impose liability upon the AV's insurer or owner.

Whether proof of culpable fault on the part of the AV is required by section 2(1) is a controversial point. Views of respondents to the Law Commission's 2018-2019 consultation on AV's were split on the issue, with some noting the confusion to which the introduction of an anthropomorphic notion of fault might lead (see the Law Commission's analysis of responses to the 2018-2019 consultation on AV's, at pages 88 to 92[7]).

For the time-being, we do no more than highlight the controversy in relation to AV fault. It is too early to tell how it might resolve. It does, however, lead to some interesting speculations as to the results of future cases. We deal with that point in the "Questions" section, below.

Finally, there is the possibility that an accident will occur, involving an AV, but that the cause will remain mysterious. This category might be described as falling within the current category of causation cases where there is "scientific uncertainty about causal mechanisms" (to quote the label put to such cases by Clerk and Lindsell on Torts). In such cases the courts lack sufficient expert evidence to distinguish between concurrent causes and have previously, in a non-AV setting, allowed causation to be proven on the basis of a material increase in the risk of injury (as in the mesothelioma cases)[8]. The findings of fact or law in such a case under the AEVA 2018 are unpredictable, but it might bring into play section 8(3)(b), among others, which provides for partial causation.

---

7 Law Commissions' full analysis of responses to the automated vehicles consultation, at https://s3-eu-west-2.amazonaws.com/lawcom-prod-storage-11jsxou24uy7q/uploads/2019/06/Automated-Vehicles-Analysis-of-Responses.pdf

8 Clerk & Lindsell on Torts (22nd edition, 2018), Chapter 2 ("Causation in Tort: General Principles"), section 2 ("Factual Causation"), Part (c) ("Scientific uncertainty about causal mechanisms").

QUESTIONS OF FUTURE COURT APPROACHES TO "ACCIDENT CAUSED BY AN AUTOMATED VEHICLE" UNDER AEVA 2018

How the AEVA 2018 will work in practice is as yet unknown. The answers to the following questions will depend upon the facts of the cases before the courts and upon the contents of law then available (regulations and case-law). So we detect the questions, but intentionally do not offer the answers here.

- How would the AEVA 2018 deal with an accident in which initial manoeuvre(s) by one or more vehicles lead other vehicles into further collisions (eg. a "concertina effect")? Section 8(3)(a) allows such a sequence of accidents ("two or more causally related accidents") to fall within the statutory liability as a matter of principle, and section 8(3)(b) (which allows an accident "partly caused by" an AV) complements that. But would the AEVA apply primary liability to an insurer or owner of an AV, involved in a concertina of causally-related accidents, even where the AV was not the primary cause of those accidents but just one of the partially-causative pieces in the concertina? Is the damage ("that damage": 2(1)) for which the insurer is liable under the Act the damage attributable to "the accident" (of which the insured AV might only be a partial cause), or only the part of the damage which the insurer might argue the AV has caused? Does the AEVA impose such a full liability on the insurer, then allow the AV insurer to use its remedy of contribution against the other culpable vehicles (if AVs under the Act) and/or human drivers or riders (of non-AVs)?

- Could an accident occur in which an automated vehicle drove carefully and well (by an objective assessment of its own driving), but in a way which caused (or was part of the cause of) an accident when it drove among less careful, human drivers? In that situation, would the AV cause the accident, so leaving its insurer primarily liable under section 2 AEVA 2018, but able to assert contributory negligence against the claimant, and to seek contribution from others?

- Might the terms of the insurance of the AV, defining "accident" (for example) be relevant to the section 2 AEVA causation issue?

- Might an accident occur in which AVs were involved, but in which there was (even on expert evidence relating to all the drivers and vehicles, AV and otherwise) no discernible cause? How would the court determine an AEVA claim? What are the limits of inference and of scientific evidence in such a case?

- To what questions would expert evidence go? Would the investigation of an AV accident need to go further than externally-apparent causes, and into the design of software in the vehicle (even if no software alteration was suspected)? In some cases, are accident reconstruction experts to be complemented or replaced by experts in fields including algorithms?

- How do these, and other AEVA 2018 considerations affect the ways in which the civil courts currently receive evidence? Will the current civil procedure in relation to electronic disclosure (Practice Direction 31B to Part 31 of the Civil Procedure Rules 1998) be revised? Will e-documents be better understood by courts in their original e-versions, on screen, rather than in printed form? How will those factors change civil trials?

CONCLUSION

Causation issues in road traffic accident claims will change with the activation of the AEVA 2018, as will the ways in which those issues are tried by the courts. The list above is bound to change, and to grow.

More, as yet unimagined causation issues will arise. Those cases will involve people – including the judges who decide them - who travel in robotically-driven vehicles.

We can map the results to some extent, but they are unpredictable. As Hart and Honoré put it, the task "which the courts discharge when they determine in concrete cases the proper limits or scope of general rules" is an "often creative function"[9].

SUMMARY OF POINTS ON CAUSATION IN THE AEVA 2018

Artificial intelligence opens a very wide debate, philosophically, scientifically and legally, in relation to determining the "cause" of accidents in which it is involved.
It is arguable, though controversial, that the AEVA 2018 has been framed to remove the burden of proving AV "fault", in the sense of humanlike blameworthiness, in an attempt to simplify that debate.
Causation remains, nevertheless, a legally difficult field, in which future approaches are difficult to predict.
Causation under the AEVA 2018 is likely to be a litigious issue, in part due to the strength of the AEVA remedy for a claimant.
Causation under the AEVA 2018 might deliver surprising results (including a change in the situations in which motor insurers have previously been held liable).
The evidence required to try AEVA 2018 claims (including expert and e-documents) and the trial process are likely to require reform.

---

9 Hart and Honoré (above) at p.4.

CHAPTER FOUR
CONTRIBUTORY NEGLIGENCE IN THE AEVA 2018

INTRODUCTION

In the previous chapter, we noted that the AEVA 2018 defines the question of causation primarily by reference to whether the "accident" was caused by the insured automated vehicle, when driving itself on a road or other public place in Great Britain (section 2(1)(a)). We suggested that, in asking that question, the 2018 Act seeks to avoid philosophically difficult questions about the blameworthiness of a machine.

The Act cannot, however, avoid that difficulty when comparing the conduct of a claimant driver with the actions of an automated vehicle, for the purpose of assessing comparative fault when both claimant and robot have contributed causatively to the same accident.

That comparison between faults of a human claimant (whether driver, rider, pedestrian or other) on the one part and robot vehicle on the other is essential to the exercise of assessing contributory negligence. So the problem looms again: how to assess the fault of a machine?

How the 2018 Act tackles that problem is the subject of this Chapter.

OBJECTIVES OF THIS CHAPTER

This chapter:

- Sets out the provisions of AEVA 2018 dealing with contributory negligence of a human driver, in an AEVA claim arising from an accident partly caused by an automated vehicle

- Discusses the difficulties in the language of the AEVA 2018, summarises the controversy and suggests a possible interpretation.

- Asks questions as to how the courts might deal with other contributory negligence issues in future, hypothetical AV cases.

THE PROVISIONS OF AEVA 2018 RELATING TO CONTRIBUTORY NEGLIGENCE

Section 3 AEVA 2018, "Contributory Negligence etc", provides as follows:

"3(1) Where—

(a) an insurer or vehicle owner is liable under section 2 to a person ("the injured party") in respect of an accident, and

(b) the accident, or the damage resulting from it, was to any extent caused by the injured party,

the amount of the liability is subject to whatever reduction under the Law Reform (Contributory Negligence) Act 1945 would apply to a claim in respect of the accident brought by the injured party against a person other than the insurer or vehicle owner.

(2) The insurer or owner of an automated vehicle is not liable under section 2 to the person in charge of the vehicle where the accident that it caused was wholly due to the person's negligence in allowing the vehicle to begin driving itself when it was not appropriate to do so."

Section 6 AEVA 2018, "Application of Enactments", provides at **subsection 6(3)** that:

"For the purposes of section 3(1), the Law Reform (Contributory Negligence) Act 1945 and section 5 of the Fatal Accidents Act 1976

(contributory negligence) have effect as if the behaviour of the automated vehicle were the fault of the person made liable for the damage by section 2 of this Act."

Section 1 of the Law Reform (Contributory Negligence) Act 1945 ("Apportionment of liability in case of contributory negligence") provides in **subsection (1)1** (as far as is relevant to the points made in this chapter) as follows:

"Where any person suffers damage as the result partly of his own fault and partly of the fault of any other person or persons, a claim in respect of that damage shall not be defeated by reason of the fault of the person suffering the damage, but the damages recoverable in respect thereof shall be reduced to such extent as the court thinks just and equitable having regard to the claimant's share in the responsibility for the damage ..."

Section 5 ("Contributory Negligence") of the Fatal Accidents Act 1976 provides as follows:

"Where any person dies as the result partly of his own fault and partly of the fault of any other person or persons, and accordingly if an action were brought for the benefit of the estate under the Law Reform (Miscellaneous Provisions) Act 1934 the damages recoverable would be reduced under section 1(1) of the Law Reform (Contributory Negligence) Act 1945, any damages recoverable in an action ... under this Act shall be reduced to a proportionate extent.'

CONTRIBUTORY NEGLIGENCE UNDER THE AEVA 2018: THE CONTROVERSY

The controversy has arisen mainly from the language of the AEVA 2018 in relation to contributory negligence. The Law Commissions, in their joint preliminary consultation paper on automated vehicles (November 2018), wrote that:

> *"The combined effect of these provisions can be quite difficult to follow"*[1]

The responses to the Law Commissions' consultation, published in 2019, did not establish a consensus, though they did offer some explanations. The writer was among the respondents to that consultation[2]. The following is his approach. It is not definitive: as repeated elsewhere in this Book, the AEVA 2018 is (at the time of writing) not in force, no regulations under the Act have been published and its concepts are untested in court.

The Problem

The problem is in the language of the 2018 Act. It might also lie in the sequence of the provisions of the Act (which arguably do not start with the first part of the answer).

Underlying that is the courts' main factual problem, which we have already discussed in relation to causation: road traffic accidents usually require a comparison of the faults of several actors. Until now, those actors have all been humans – drivers, riders, pedestrians etc. But how are the courts to assess the fault of an artificially intelligent machine?

Might the answer to the wording of these AEVA 2018 provisions lie, again, in that central problem?

---

1 Law Commissions' Joint Preliminary Consultation Paper on Automated Vehicles (CP240, DP 166, 8 November 2018) at p.107 §6.36, at https://s3-eu-west-2.amazonaws.com/lawcom-prod-storage-11jsxou24uy7q/uploads/2018/11/6.5066_LC_AV-Consultation-Paper-5-November_061118_WEB-1.pdf

2 The responses are all available on the Law Commission of England and Wales' Automated Vehicles website, at https://www.lawcom.gov.uk/project/automated-vehicles/. The Law Commission's analysis of the responses on causation and contributory negligence are at pages 85 to 91 of its Analysis document: https://s3-eu-west-2.amazonaws.com/lawcom-prod-storage-11jsxou24uy7q/uploads/2019/06/Automated-Vehicles-Analysis-of-Responses.pdf

The Difficult Language of the AEVA 2018

The difficulty arises particularly from the final phrase in subsection 3(1):

> "… the amount of the liability is subject to whatever reduction under the Law Reform (Contributory Negligence) Act 1945 would apply to a claim in respect of the accident brought by the injured party against a person other than the insurer or vehicle owner."

Why does that subsection introduce this new character, "a person other than the insurer or vehicle owner"? Who is that person? Why is he in the Act?

Shuffling the Statutory Cards

One way to approach the problem is to step away from it, for a moment, and consider the other provisions set out above, relating to contributory negligence.

What do the other provisions do? We can re-shuffle the deck of cards. We can start with the earliest.

- Section 1 of the 1945 Act. As the title of the Act says, this was a Law Reform Act. The reform was to change the previous position, described in section 1: "*Where any person suffers damage as the result partly of his own fault and partly of the fault of any other person or persons, a claim in respect of that damage shall not be defeated by reason of the fault of the person suffering the damage*". One of the main reasons for that reform was accidents involving motor cars. Those accidents tended to happen through the faults of more than one actor. That was a change in the way in which the majority of plaintiffs[3] had been injured. It was a technological change which required the reform of tort law. As Professor Paul Mitchell described, in his book "A History of

---

3 The former term for claimants.

Tort Law 1900-1950" (Cambridge, 2015), in 1934 the Law Revision Committee had invoked "the position which frequently arises when the plaintiff sustains a single damage from the combined negligence of two motor car drivers" as a reason for allowing apportionment of responsibility between tortfeasors (by apportionment and contribution, the AEVA 2018 remedy for which we consider in a later chapter). Motor technology similarly propelled "the need for a common-sense approach to causation" in contributory negligence[4]. So the comparison of different actors' faults, and the attribution of fault between them, stems from this current Act (the 1945 Act). It is echoed in the 1976 Act relating to Fatal Accidents. Both Acts speak of a "person".

- Section 5 ("Contributory Negligence") of the Fatal Accidents Act 1976 provides for the death of a human person (or at least for the first person in the opening phrase): "*Where any person dies as the result partly of his own fault and partly of the fault of any other person or persons …*".

At this stage we deal a new card into the deck:

- The Interpretation Act 1978, Schedule 1, defines a person as including "a *body of persons corporate or incorporate*". So that word does not seem conclusive of human identity. Although this appears in the 1978 Act, it is a reiteration of an earlier (1889) definition.

- Then we come back to the AEVA 2018:

---

4 See Mitchell at chapters 11 and 13. The first quotation is from the 1934 Law Revision Committee, at Mitchell page 278, the second a quotation of Professor Mitchell, at page 314. The Law Revision Committee was the precursor of the Law Commission (the politics of its and the LRC's establishment are described fascinatingly by Professor Mitchell).

- Section 3(1) AEVA 2018 ("Contributory Negligence etc") provides for that extra, notional character ("a person other than the insurer or vehicle owner"):

    "Where—

    (a) an insurer or vehicle owner is liable under section 2 to a person ("the injured party") in respect of an accident, and

    (b) the accident, or the damage resulting from it, was to any extent caused by the injured party,

    the amount of the liability is subject to whatever reduction under the Law Reform (Contributory Negligence) Act 1945 would apply to a claim in respect of the accident brought by the injured party against a person other than the insurer or vehicle owner."

- Section 6(3) AEVA 2018, (in "Application of Enactments") provides that:

    "For the purposes of section 3(1), the Law Reform (Contributory Negligence) Act 1945 and section 5 of the Fatal Accidents Act 1976 (contributory negligence) have effect as if the behaviour of the automated vehicle were the fault of the person made liable for the damage by section 2 of this Act."

How do these sections work?

That was the question posed by the Law Commissions' consultation.

In the writer's view, the answer might be as follows:

- The sections from the AEVA 2018 set out Parliament's approach to assessing contributory negligence in an AEVA 2018 claim. The particularity of the AEVA 2018 scheme is a notable feature of the Act.

- The AEVA 2018 adopts the methods of other legislation in this respect (the 1945 Act and the 1976 Fatal Accidents Act). It is noteworthy that the AEVA 2018 does so particularly in relation to contributory negligence; it does not do so in relation to the separate, though thematically-linked remedy of contribution (section 6(5) AEVA 2018 excludes a right of contribution under the Civil Liability (Contribution) Act 1978, as we will discuss in Chapter 6).

- The AEVA 2018 makes the insurer or owner, who is liable under section 2, vicariously liable for the conduct of the automated vehicle. That is the effect of section 6(3) (which gives effect in an AEVA 2018 claim to the 1945 and 1976 Acts "*as if the behaviour of the automated vehicle were the fault of the person made liable for the damage by section 2 of this Act*").

- In relation to the standard of care applicable to the automated vehicle, in the assessment of its fault in comparison to that of the contributorily negligent claimant, the AEVA 2018 introduces a legal fiction. That is the introduction of an extra comparator ("a person other than the insurer or vehicle owner"), as a substitute for the otherwise available comparators on the AV's side (the AV itself, or its insurer or owner).

If that is correct, why does the 2018 Act introduce that extra comparator?

A Possible Explanation for the Language of section 3(1) AEVA 2018

In the writer's view, the reason might be the same as that raised in the causation chapter: to avoid the philosophical difficulty of comparing human and robotic behaviour, and the underlying reasoning of both.

That comparison between human and robotic reasoning would present a formidable exercise, if (contrary to the writer's view) it was what the legislature intended the courts to undertake when adjudicating AEVA 2018 claims.

It is a comparison which seems not to reflect the actuality of artificially-intelligent systems. To quote the mathematician Dr Hannah Fry, in her book about algorithms and artificial intelligence: "Hello World: How to be Human in the Age of the Machine" (Penguin, 2018) at page 13:

> "The downside is that if you let a machine figure out the solution for itself, the route it takes to get there often won't make a lot of sense to a human observer".

That might be the problem at which the legal fiction in section 3(1) – the introduction of a fictional human ("person") to stand in the place of the AV for the purpose of the contributory negligence assessment – is aimed.

CONCLUSION

It must be acknowledged that the answer is not clear either way. The parliamentary intention behind the framing of contributory negligence in the 2018 Act was stated, by the government, to be that the Bill (as it then was) "mirrors existing processes as closely as possible without making complex legislative changes to the legal framework …" (letter from John Hayes MP dated 15 November 2017[5]), though that assertion lacks binding legal effect.

It might also be argued that the word "person", as defined by the Interpretation Act 1978, permits of non-human identity (including bodies corporate), so is not conclusive of a human comparator (though the counter-argument might be that legislation concerning comparative fault before Artificial Intelligence did not contemplate independent

---

5 In "Will Write Letters" at https://services.parliament.uk/bills/2017-19/automatedandelectricvehicles/documents.html

robots, and that the meaning of "person other than the insurer or vehicle owner" in section 3(1) is tolerably clear as meaning a human person).

In the writer's view, the legal fiction of a human comparator seems, for the time-being (and in the absence of regulations and case law on AEVA 2018), a persuasive interpretation of section 3 AEVA 2018.

If so, it has been argued that such a standard would be unacceptably anthropomorphic[6]. But, if that were correct, the next question would be "How are judges to compare human with robotic reasoning?". And there, the problem highlighted by Dr Hannah Fry returns: the comparison between human and AI reasoning is not easily made. As the writer argued in response to the Law Commission's 2018-2019 consultation, it is:

> "a comparative question which has vexed authors and scientists[7] and which would, at least on present scientific knowledge, probably also ensnare the courts if engaged in the essentially comparative[8] debate on contributory negligence"[9]

OTHER QUESTIONS OF FUTURE COURT APPROACHES TO CONTRIBUTORY NEGLIGENCE UNDER AEVA 2018

We reiterate that how the AEVA 2018 will work in practice is as yet unknown. The answers to the questions posed in this chapter (and

---

6 See the Law Commission's full analysis of the responses to the consultation at pages 85 to 91 (endnotes above).

7 Eg. "Do Androids Dream of Electric Sheep?" by the novelist Philip K Dick (1968) and "Life 3.0" by the physicist Max Tegmark (2017).

8 See van Dongen, E.G.D. and Verdam, H.P., 2016. The Development of the Concept of Contributory Negligence in English Common Law, *Utrecht Law Review*, 12(1), pp.61-74. DOI: http://doi.org/10.18352/ulr.326

9 See all responses to the Law Commissions' 2018-2019 consultation, including the writer's, at https://www.lawcom.gov.uk/draft-responses-to-the-automated-vehicles-consultation-2018-19/

others unimagined by us) will depend upon the facts of the cases before the courts and upon the contents of law then available (including regulations and case-law). So we do not offer definitive answers here.

But the other questions might include:

- How will courts assess human contributory negligence, when it is alleged (by the insurer, in a sense on behalf of the AV) that the human person-in-charge failed to take back control of the AV from the automated system, in circumstances where the human ought to have done so (as discussed in Chapter 1)? This situation is one of the reasons that the AEVA 2018 avoids (it is said) coverage of Level 3 vehicles, "the murky middle" of liability (as the Law Commission describes it) in which control is shared to a greater degree between human and AV than at the subsequent levels 4 and 5 (the latter implying totally automated driving). But this issue will still arise under the AEVA 2018, when dealing with an injurious accident caused, or partly caused, by a Level 4 AV. The issue of human intervention with the automated control of a vehicle has been in the news recently, for example in relation to human pilots taking back control of automated aircraft systems. It is likely to arise in AEVA 2018 cases involving automated road vehicles. Where the boundaries of reasonable and negligent human behaviour will lie is untested, and will depend upon the facts and evidence (and especially expert evidence) in each case.

- Is 100% contributory negligence possible under the AEVA 2018? It is made expressly possible in a particular situation, by section 3(2) of the Act (above): "The insurer or owner of an automated vehicle is not liable under section 2 to the person in charge of the vehicle where the accident that it caused was wholly due to the person's negligence in allowing the vehicle to begin driving itself when it was not appropriate to do so." That is the situation where fault is likely to be so difficult to assess (see the point above). Will 100% contributory negligence be possible in AV accident situations other than that? The facts have not yet arisen.

SUMMARY OF POINTS ON CONTRIBUTORY NEGLIGENCE IN THE AEVA 2018

The language of the AEVA 2018 relating to contributory negligence is difficult
The linguistic complexity might be an attempt to simplify the assessment of comparative fault in AEVA claims, by removing the need to compare human with robot behaviour, and substituting a legal fiction of human behaviour on the AV side.
If so, that is controversial.
Contributory negligence will pose difficult questions of law and fact, for example in cases where the human driver is criticised for insufficient care in taking back control of a Level 4 vehicle.
100% contributory negligence is possible under the Act, in the situation expressly legislated in section 3(2). Whether it will be more widely available is unpredictable.

CHAPTER FIVE
EXCLUSIONS AND RESTRICTIONS IN THE AEVA 2018

INTRODUCTION

The strong remedy under the AEVA 2018 (direct liability of an insurer or owner for damage suffered by a person in an accident caused by an insured automated vehicle when driving itself on a road or other public place in Great Britain) is excluded – or restricted – in a number of circumstances.

OBJECTIVES OF THIS CHAPTER

This chapter describes:

- the situations in which the remedy of direct liability of the insurer or owner for an AV accident (section 2 of the 2018 Act) is excluded or restricted

- which of those exclusions or restrictions arise by implication (eg. the uninsured AV, which the Act does not cover) and which are set out expressly in provisions of the 2018 Act

- some potential controversies arising from those exclusions and restrictions

We cannot list all of the potential circumstances in which the 2018 Act will not apply or will be restricted in its application. Regulations under the Act, yet to be published, might provide greater clarity.

Another reason is that British motor insurance law branches significantly from European Union law. EU law is due to cease to apply in the UK as a result of the UK's withdrawal process from the EU (which is still in process at the time of writing). But the extent to which British motor insurance law has followed the European path is likely to leave a strong European influence in future UK motor insurance law, as well as sudden divergences.

Furthermore, the bringing into force of the AEVA 2018 (when it happens) will be a very significant and new feature of British motor insurance law.

Yet another reason why we cannot provide a comprehensive list of exceptions to the AEVA 2018 is, of course, the unknown effects of automated vehicle technology in practice.

That is not to say that there are few apparent exceptions and restrictions to liability under the Act as it stands. There are several.

THE PROVISIONS OF AEVA 2018 EXCLUDING OR RESTRICTING ITS APPLICATION

The 2018 Act provides for some exceptions explicitly and some by implication.

Implied Exceptions and Restrictions

The implied exceptions and restrictions are as follows:

- Section 2 ("Liability of insurers etc where accident caused by an automated vehicle") does not provide any remedy to a claimant where the automated vehicle is uninsured or otherwise secured against third-party risks. The compensation of victims of accidents attributable to the acts of uninsured or untraced drivers is not currently on a statutory footing, but is provided by voluntary agreements between government and the insurance industry (administered by the Motor Insurers' Bureau or MIB). But – as Chapter 8 – "Uninsured Vehicles" describes – no position has yet been reached in relation to the compensation of victims of accidents involving uninsured AVs.

- The AEVA 2018 has been said not to apply to vehicles described in the Society of Automotive Engineers' (SAE) definition as "Level 3". Level 3 is, broadly put, the first level of advanced driving systems, in which control can be ceded in some circumstances to the car itself. But the human driver must always be supervising the car's actions and ready to take back control of the car as soon as the need arises. Where this need to intervene will arise in practice, and when a human driver should resume control, are matters untested in law (the Law Commission refers to it as "the murky middle" of AV law). So to insurers this is a risk lacking a reasonable means of assessment, and an unattractive situation to insure (see Chapter 1).

Express Exceptions and Restrictions

The express list of exceptions and restrictions on the face of the AEVA 2018, absent regulations, is as follows:

- "Damage" suffered by an insured person or any other person, for which an insurer or owner is liable under Part 1 of the 2018 Act, is given a restricted meaning. "Damage" is defined by section 2(3) AEVA 2018 as meaning:

    "… death or personal injury, and any damage to property other than—

    (a) the automated vehicle,

    (b) goods carried for hire or reward in or on that vehicle or in or on any trailer (whether or not coupled) drawn by it, or

    (c) property in the custody, or under the control, of—

> *(i) the insured person (where subsection (1) applies), or*
>
> *(ii) the person in charge of the automated vehicle at the time of the accident (where subsection (2) applies)."*

- Section 2(4) limits the amount of liability for damage to property arising out of any one accident involving an automated vehicle to the amount for the time being specified in section 145(4)(b) of the Road Traffic Act 1988 (the limit on compulsory insurance for property damage).

- Contributory negligence, if found, applies to reduce the amount of the liability (sections 2(5) and 3(1)).

- Section 3(2) of the AEVA 2018 prescribes one type of contributory negligence which operates to extinguish AEVA 2018 liability completely:

  > *"The insurer or owner of an automated vehicle is not liable under section 2 to the person in charge of the vehicle where the accident that it caused was wholly due to the person's negligence in allowing the vehicle to begin driving itself when it was not appropriate to do so."*

- The AEVA 2018 allows insurers of AVs to exclude or limit their liability under section 2(1) of the Act in the terms of the insurance policy, but only in relation to two particular types of failure which cause an accident, both relating to the software of the vehicle (section 4: "Accident resulting from unauthorised software alterations or failure to update software").

  Those two software restrictions are particular in their terms, and section 2(6) (in the "Liability ..." section) provides that *"Except as provided by section 4, liability under this section may not be limited or excluded by a term of an insurance policy or in any other way"*.

Section 4(1) provides that:

"An insurance policy in respect of an automated vehicle may exclude or limit the insurer's liability under section 2(1) for damage suffered by an insured person arising from an accident occurring as a direct result of—

(a) software alterations made by the insured person, or with the insured person's knowledge, that are prohibited under the policy, or

(b) a failure to install safety-critical software updates that the insured person knows, or ought reasonably to know, are safety-critical."

The terms "software alterations", "software updates" and "safety critical" in section 4 are defined by section 4(6). Software "updates" and "alterations" "…in relation to an automated vehicle, mean (respectively) alterations and updates to the vehicle's software" (section 4(6(a)). "Software updates are "safety-critical" if it would be unsafe to use the vehicle in question without the updates being installed" (section 4(6)(b)).

The insurer has a right of recovery of an amount it pays under the Act, from the person who made the prohibited alteration or who failed to install safety-critical updates, in contravention of the terms of the insurance policy. The insurer's right of recovery for those prescribed software-related breaches of the policy is set out in section 4(4):

"If the accident occurred as a direct result of—

(a) software alterations made by an insured person, or with an insured person's knowledge, that were prohibited under the policy, or

> *(b) a failure to install safety-critical software updates that an insured person knew, or ought reasonably to have known, were safety-critical,*
>
> *the amount paid by the insurer is recoverable from that person to the extent provided for by the policy."*

There is a causative relationship between the prohibited software act/omission and the occurrence of the accident. But the language of causation in this section ("*the accident occurred as a direct result of software alterations…*": underlining added) is different to that used in section 2 (liability) and section 3 (contributory negligence), both of which use the phrase "caused by". Section 4(4) thus avoids that legally complex word (see Hart and Honoré, discussed in Chapter 3): "causation". This arguably puts a more demanding burden upon the person (inevitably the insurer) seeking to rely upon the exclusion.

The insurer's right of recovery for those software-related breaches of the policy is limited further against "an insured person who is not the holder of the policy", in that the insured person who is not the policyholder must be shown to have known "at the time of the accident" that the software alterations were prohibited under the policy (section 4(5)).

DISCUSSION OF THE EXCLUSIONS AND RESTRICTIONS OF LIABILITY UNDER AEVA 2018

There are, notably, fewer exceptions available to the insurer under the AEVA 2018 than there are in relation to conventional vehicles under the Road Traffic Act 1988 (compare the AEVA 2018 exclusions with the RTA 1988 regime, which allows both a longer list of exceptions than the AEVA 2018, as well as two exceptions - non-insurance for employees and deposit of security as an alternative to insurance - which the Schedule to the AEVA 2018 expressly removes from the RTA 1988 by amendment – see the "Overview of the AEVA" chapter).

The express exclusions and restrictions of liability under the 2018 Act are clear on their own terms. The implied exclusions are less clear. We pick out two. The lack of remedy for uninsured AVs is discussed in chapter 8. The question of the level(s) of automation to which the AEVA 2018 applies we discuss here.

The Implied Exclusion of Level 3 (and possibly Level 4) Vehicles

The AEVA 2018 is said by the government not to apply to vehicles described in the Society of Automotive Engineers (SAE) definition as "Level 3".

That exclusion is not set out expressly in the Act. The question of the extent to which the AEVA 2018 applies to different levels of highly automated vehicles arises (as already discussed in Chapters 1 and 2) from the definition given to "automated vehicle" by sections 1 and 8.

Understanding the definition of "automated vehicle" in the Act, and the reasons for that definition, requires some background information.

Level 3 is, broadly put, the first level of advanced driving systems, in which control can be ceded in some circumstances to the car itself. Beyond Level 3, the extent of automation increases until at Level 5 the vehicle is fully automated, i.e. it never requires a human driver.

Whether or not manufacturers of automated vehicles have yet reached Level 3 is itself unclear. Commercial considerations apply (including, on the one hand, several manufacturers' aim to be the first to market a fully self-driving car and, on the other hand, their desire to guard commercial secrets). Publicity likes to advertise the aspiration (sometimes as if it has already been achieved) while commercial sensitivity prefers to hide the detail.

What is clear is that, during the parliamentary course of the Bill which in July 2018 became the Automated Vehicles Act 2018, parliamentarians felt some concern as to Level 3.

The reason for their concern was the risk posed by Level 3. The particular, potential hazard of Level 3 is that, while the car is able to take control of driving in certain circumstances (a motorway is a common example), the human driver must always be supervising the car's actions and ready to take back control of the car as soon as the need arises.

Many have pointed out the tension in that situation: the human driver is, simultaneously, being told (e.g. by AV manufacturers) that he or she can relax, but also (eg. by public authority) that he or she must also be able to intervene, to take back control of the vehicle, at a moment's notice.

Furthermore, that lack of flow (to coin a phrase) in the human driver's mental state might lead to a slower, intervening response, than the response of a human driver who knows that it is their task to concentrate continuously.

Where this need to intervene will arise in practice, and when a human driver should resume control, are matters untested in road transport law (the Law Commission refers to it as "the murky middle" of legal prediction).

So to insurers, Level 3 automated vehicles present risks lacking a reasonable means of assessment, and are therefore a most unattractive situation to insure. The Association of British Insurers, jointly with Thatcham Research, went as far as stating (in their response to the Law Commissions' 2018-2019 consultation on AVs) that Level 3 systems themselves "should be discouraged as they are likely to confuse drivers, who may over-rely on the system when it is not appropriate to do so"[1] (and see the discussion of intervention in Chapter 1).

But, will that opposition to Level 3 – apparently on the basis that it is too hazardous a technology even to be released to the public – succeed

---

1 ABI and Thatcham Research joint response to the Law Commission Consultation, page 4 para.7, at https://s3-eu-west-2.amazonaws.com/lawcom-prod-storage-11jsx-ou24uy7q/uploads/2019/06/AV001-ABI-and-Thatcham-Research-joint-response.pdf

in blocking the development of Level 3 vehicles? That seems unlikely. Level 3 is, logically, a likely stage of development in AVs. Even in the weeks before the writing of this Chapter, there was news (the authority of which the writers cannot endorse) that fully self-driving software has become available to download to a particular vehicle.

The question, therefore, might arise as to whether or not the AEVA 2018 might in fact cover a "Level 3" automated vehicle.

If so, it might be observed that the AEVA 2018 does not in fact define automated vehicles by reference to the SAE levels. Instead, it sets its own statutory definition. Under section 1, the Secretary of State for Transport (SoST) has a duty to keep a list of all automated vehicles lawfully allowed to drive themselves on roads or public places in the UK. Under that section it is a question for SoST as to whether the vehicle in question is, in the opinion of SoST, within that definition.

To remind ourselves, the definition of an "automated vehicle" appears in the following sections of the 2018 Act:

"Section 1: Listing of automated vehicles by the Secretary of State

(1) The Secretary of State must prepare, and keep up to date, a list of all motor vehicles that—

(a) are in the Secretary of State's opinion designed or adapted to be capable, in at least some circumstances or situations, of safely driving themselves, and

(b) may lawfully be used when driving themselves, in at least some circumstances or situations, on roads or other public places in Great Britain.

(2) The list may identify vehicles—

(a) by type,

(b) by reference to information recorded in a registration document issued under regulations made under section 22 of the Vehicle Excise and Registration Act 1994, or

(c) in some other way.

(3) The Secretary of State must publish the list when it is first prepared and each time it is revised.

(4) In this Part "automated vehicle" means a vehicle listed under this section".

"Section 8: Interpretation

(1) For the purposes of this Part—

(a) a vehicle is "driving itself" if it is operating in a mode in which it is not being controlled, and does not need to be monitored, by an individual;

...

(2) In this Part—

"automated vehicle" has the meaning given by section 1(4)"

The phrase which (it was said by the British government) would prevent the AEVA 2018 applying to an accident involving an SAE Level 3 automated vehicle is the last phrase in subsection 8(1)(a): "and does not need to be monitored" by an individual.

That would exclude SAE Level 3 vehicles, which need to be monitored by a User in Charge at all times.

That wording might also, at first sight, have excluded SAE Level 4 vehicles which, to a lesser extent than Level 3, still require a degree of human supervision. That would lead to a curious result: that the AEVA

2018 would not apply at all, until automated vehicles could operate at Level 5, in other words fully autonomously and without any need for human supervision.

If that were so, the AEVA 2018 would sit on the statute book, inactive, for some years before the arrival of Level 5, fully autonomous vehicles on British roads. And victims of any accidents involving insured Level 3 and 4 automated vehicles would not have any direct remedy against the insurer under the AEVA 2018.

The question of which levels of higher vehicle automation the Bill was intended to regulate was raised with the government, in parliament, before the Bill became law. The government's response is contained in a letter from Baroness Sugg to Baroness Randerson dated 13 March 2018, which appears on the UK parliament's webpage relating to the AEVA 2018[2]. The material part of Baroness Sugg's letter is as follows:

> *"In my closing remarks addressing the question you raised about whether the Bill's provisions cover Level 3 vehicles, I stated that the measures in Part 1 of the Bill would relate to Level 5 vehicles."*

> *"The Bill is intended to cover the use of highly automated and fully automated vehicles which are broadly equivalent to Level 4 and 5 as defined by the Society of Automotive Engineers, and not just Level 5 as I stated. To reiterate my response in the debate, and as set out in the consultation documents, the Bill does not cover conditionally automated (broadly equivalent to Level 3) vehicles."*

> *"We have not referred directly to the SAE levels in the Bill text as these are industry standards, and therefore could change over time. In addition, as the levels of automation outline the broad capability, rather than a specific function, this makes it challenging for them to be type approved or standardised. As the technology is still in development, regulating or defining standards now is likely to create, rather than remove, impediments to innovation. This is why we have*

---

2 The letter is at http://data.parliament.uk/DepositedPapers/files/DEP2018-0264/Baroness_Sugg_-_Baroness_Randerson_AEV_Bill_2nd_reading.pdf

used the definition within the Bill of a vehicle 'safely driving itself' and incorporated within it the point that the vehicle 'does not need to be monitored' by an individual."

So the government's answer was: The Automated and Electric Vehicles Bill applies to Levels 4 and 5, but not Level 3. That was despite the wording of the Bill (later enacted) as to residual human control (which would seem to apply it only to Level 5, which seemed to have been Baroness Suggs' understanding when asked in parliament).

So there had been some understandable confusion in parliament as to the effect of the wording of the Bill: the AEV Bill's remedy seemed initially to have contracted to Level 5 alone (with the inclusion of the phrase "and does not need to be monitored by") but then expanded, albeit slightly, to include Level 4. But (in the government's reasoning at that stage) it had not expanded downwards so far as to include Level 3 (as might have been the case before the inclusion of the "… monitored …" phrase).

In a subsequent letter, dated 13 April 2018, Baroness Sugg on behalf of the government commented further on the "Inclusion of an additional definition for 'control'", as follows[3]:

> "*The term 'controlled' is used in the Bill in the definition 'driving itself' – a vehicle is 'driving itself' if it is operating in a mode in which 'it is not being controlled, and does not need to be monitored by, an individual'. We consider the concept here is clear from these words. If a more detailed description were used there is a risk of the term inadvertently departing from the policy intention, particularly given that we are legislating to cover future technology.*"

> "*We do not, therefore, feel that it is necessary to provide a separate definition of control within the AEV Bill.*"

---

3 At the second page of the letter, which is at http://data.parliament.uk/Deposited-Papers/files/DEP2018-0391/Baroness_Sugg_letter_-_terminology_in_the_AEV_Bill.pdf

Whether "the concept is clear" as to which levels of higher vehicle automation the AEVA 2018 will govern, when brought into force by regulations, will depend upon the content of those regulations, as well as upon the list by then written by the Secretary of State for Transport.

It might be that regulations, the Secretary of State's list of AV's or other government documents will tell us the answer. For example, it might be that the Secretary of State will consider an automated vehicle to fall within the AEVA 2018, even if others (perhaps including the SAE) consider it to be a Level 3 vehicle. A contest upon that definitive issue in court would add considerably to the costs of individual claims for damages in AV cases, by opening up a new issue for expert evidence.

But if those future government documents add nothing to the language of the Act, the puzzle will remain: in the face of that definition of an automated vehicle, how will the victim of an AV accident know whether or not they can sue the insurer directly under the Act?

SUMMARY OF POINTS ON EXCLUSIONS AND RESTRICTIONS TO LIABILITY UNDER AEVA 2018

The AEVA 2018 is limited in its application both expressly, by its own provisions, and implicitly, by reference to external factors
The AEVA 2018 might be further affected by circumstances and laws not yet in effect (including regulations and the SoST's list of AVs yet to be published under the 2018 Act, the way in which AV technology will act when it is released and the legal effects of withdrawal from the EU, particularly in relation to motor insurance law)
The AEVA 2018 provides insurers with a limited right to avoid cover of, and to recover amounts paid by the insurer from, certain people who, in breach of the insurance policy, alter the AV's software and/or fail to install safety-critical software updates
The Act does not provide a remedy for the victim of an accident involving an uninsured or otherwise financially-secured AV
The level(s) of automation subject to the liability provision of the Act is likely to be a controversial issue

CHAPTER SIX
THE INSURER OR OWNER'S RIGHT TO CLAIM AGAINST ANOTHER RESPONSIBLE PERSON UNDER THE AEVA 2018

INTRODUCTION

The AEVA 2018 allows an insurer or owner found liable to compensate the victim of an accident caused when driving itself (section 2), the amount of which liability is fixed (section 5(2)), to claim against another responsible party.

That type of action is currently known (in non-AV cases) as a third-party claim or claim for contribution. The AEVA 2018 does not use those terms.

OBJECTIVES OF THIS CHAPTER

This chapter:

- Sets out the provisions of the AEVA 2018 relating to the liable insurer or owner's right to claim against another person responsible for the accident

- Discusses the scheme of the Act in relation to such claims by an insurer or owner against another responsible person

- Asks some questions as to how that scheme might work in practice

THE PROVISIONS OF AEVA 2018 ABOUT AN INSURER OR OWNER'S RIGHT TO CLAIM AGAINST ANOTHER RESPONSIBLE PARTY

Section 2(1) ("Liability of insurers etc where accident caused by automated vehicle", already discussed in Chapters 2 and 3) provides that:

> "*Where—*
>
> *(a) an accident is caused by an automated vehicle when driving itself on a road or other public place in Great Britain,*
>
> *(b) the vehicle is insured at the time of the accident, and*
>
> *(c) an insured person or any other person suffers damage as a result of the accident,*
>
> *the insurer is liable for that damage.*"

Section 2 has effect subject to the provisions of the Act relating to contributory negligence (section 2(5) and 3, already discussed in Chapter 4).

Section 2(7) provides that:

> "*The imposition by this section of liability on the insurer or vehicle owner does not affect any other person's liability in respect of the accident.*"

Section 5 provides (in its entirety) as follows:

> "**5 Right of insurer etc to claim against person responsible for accident**
>
> *(1) Where—*

(a) section 2 imposes on an insurer, or the owner of a vehicle, liability to a person who has suffered damage as a result of an accident ("the injured party"), and

(b) the amount of the insurer's or vehicle owner's liability to the injured party in respect of the accident (including any liability not imposed by section 2) is settled,

any other person liable to the injured party in respect of the accident is under the same liability to the insurer or vehicle owner.

(2) For the purposes of this section, the amount of the insurer's or vehicle owner's liability is settled when it is established—

(a) by a judgment or decree,

(b) by an award in arbitral proceedings or by an arbitration, or

(c) by an enforceable agreement.

(3) If the amount recovered under this section by the insurer or vehicle owner exceeds the amount which that person has agreed or been ordered to pay to the injured party (ignoring so much of either amount as represents interest), the insurer or vehicle owner is liable to the injured party for the difference.

(4) Nothing in this section allows the insurer or vehicle owner and the injured party, between them, to recover from any person more than the amount of that person's liability to the injured party.

(5) For the purposes of—

(a) section 10A of the Limitation Act 1980 (special time limit for actions by insurers etc in respect of automated vehicles), or

(b) section 18ZC of the Prescription and Limitation (Scotland) Act 1973 (actions under this section),

the right of action that an insurer or vehicle owner has by virtue of this section accrues at the time of the settlement referred to in subsection (1)(b)."

Section 6(4) ("Application of enactments") provides that:

"Liability under section 2 is treated as liability in tort or, in Scotland, delict for the purposes of any enactment conferring jurisdiction on a court with respect to any matter."

Section 6(5) provides that:

"An insurer or vehicle owner who has a right of action against a person by virtue of section 5 does not have a right to recover contribution from that person under the Civil Liability (Contribution) Act 1978 or under section 3 of the Law Reform (Miscellaneous Provisions) (Scotland) Act 1940."

The new section 10A(1) of the Limitation Act 1980 (special time limit for actions by insurers etc in respect of automated vehicles), inserted by section 20 and paragraph 9 of the Schedule to AEVA 2018, provides as follows:

*"**10A Special time limit for actions by insurers etc in respect of automated vehicles**"*

"(1) Where by virtue of section 5 of the Automated and Electric Vehicles Act 2018 an insurer or vehicle owner becomes entitled to bring an action against any person, the action shall not be brought after the expiration of two years from the date on which the right of action accrued (under subsection (5) of that section)."

DISCUSSION OF THE "OTHER RESPONSIBLE PERSON" PROVISIONS OF THE AEVA 2018

As in other chapters, we cannot (in the absence of regulations under the Act and case law interpreting it in practice) be comprehensive or certain in our views on the Act. But the following points emerge.

First, the AEVA 2018 does not speak in terms of "contribution" or "third party" claims, nor of "indemnity". Those terms might, in future case law, be used by the courts in relation to the remedy under section 5 AEVA 2018 (for example, the effects of sections 5(3) and 5(4) might echo indemnity principles preventing overcompensation). But the Act itself does not use those words. In fact, it explicitly denies an insurer or owner the remedy of contribution provided by the Civil Liability (Contribution) Act 1978[1]. Section 5 AEVA 2018 instead confers upon the insurer or owner made liable by section 2, a *"Right ... to claim against person responsible for accident"* (in the summary headline of the section).

Second, that claim (by the insurer or owner, against that other responsible person) is – like the victim's section 2 claim against the insurer or owner – another type of direct claim. The effect of section 5(1) is that where an insurer or owner has been made liable to the

---

1 In response to the Law Commission's 2018-2019 AV consultation, the writer speculated that this might be an attempt at simplification related to the criterion for entitlement to contribution in section 1 of the 1978 Act that liability be "in respect of the same damage" suffered by the victim. That criterion caused some controversy in road transport law, in relation to whether an insurer seeking contribution to credit hire charges from a third party's dilatory repairer was suing for the "same damage" (its liability to the Claimant in hire charges) as that suffered by the Claimant (damage to the car, if that is the correct comparison). The Court of Appeal had indicated, possibly *obiter dicta* (not in relation to the legal point decided), that the insurer could seek contribution to its liability to pay the hire charges from the repairer (*Clark v Ardington* [2003] QB 36 at §121). But a circuit judge of the county court subsequently held otherwise, on his different understanding that this was not the "same damage" as that suffered by the Claimant, within the 1978 Act (*Mason v TNT and Groupama* 13 April 2009, Oxford county court, LTL document no. AC0121869, 6 August 2009). But that is speculation: it might be that there are different reasons for the abandonment of the 1978 Act in the AEVA 2018 scheme.

victim under section 2, and the amount of that liability has been settled (by an amount of damages, typically), "*any other person liable to the injured party in respect of the accident is under the same liability to the insurer or vehicle owner*". So the liable insurer or owner has, by virtue of its liability to the victim, acquired a right of action against any other person liable to the victim in respect of the accident.

Third, the other responsible person's liability is not as narrow as the section 2 AEVA 2018 liability. That seems to follow, in the first place, from the wording of section 5(1) (which distinguishes between the insurer or owner's "section 2" liability in 5(1)(a) – which is tortious (6(4)) - and the liability to the victim of the other person "in respect of the accident" at the end of section 5(1), which appears not to be restricted to tortious liability). And it seems to follow from the nature of the insurer or owner's right under section 5 against another responsible person, which would only have any practical effect if that other responsible person was capable of being something other than the insurer or owner of another automated vehicle responsible for the accident (the other responsible person could be, for example, another actor in the accident – a driver of a non-AV, a pedestrian, a cyclist etc – or a person with a more distant influence, eg. a vehicle designer or manufacturer – the potential bases of whose liability are discussed in Chapter 9, "Product Liability Claims").

Fourth (and taking its lead from the third point), could the insurer/owner liable under section 2 AEVA 2018 bring a section 5 claim against another AV owner or insurer? The answer is uncertain for the reasons set out at the start of this discussion. But it might be that the section 5 claim could apply in those circumstances. Section 2(7) (above) provides that "The imposition by this section [section 2] of liability on the insurer or vehicle owner does not affect any other person's liability in respect of the accident".

That might be read broadly to include the insurer or owner of a second AV involved in the same accident. That might make good practical sense. Whether a claim by a victim against two insurers/owners of two automated vehicles would result in two sets of proceedings directed to be heard on a single occasion (as often happens now, in conventional vehicle cases) or otherwise, by consecutive actions, would depend upon the facts and economics of individual cases.

It is notable that, in a multi-AV accident claim, the AEVA 2018 allows a claimant victim the choice to sue just one insurer/owner defendant and to allow that defendant to decide whether and how to pursue its own remedy against another potentially liable AV. There is also the difference in limitation regimes between sections 2 and 5 AEVA 2018 to be taken into account.

Those will all be practical considerations for parties, their legal representatives and the courts in future cases.

SUMMARY OF POINTS ON THE INSURER OR OWNER'S CLAIM AGAINST OTHER RESPONSIBLE PARTY UNDER AEVA 2018

The AEVA 2018 provides a liable insurer or owner with a right to claim against another responsible person
That right to claim against another party is not defined as "contribution", excludes the 1978 Act remedy and might be broader in practice than current contribution proceedings
How the insurer or owner's right to claim will interact with other causes of action (both under section 2 AEVA 2018 and otherwise) in future claims arising from multi-AV accidents is unknown, but might be another explanation for the breadth of the AEVA 2018 scheme.

PART THREE

AUTOMATED VEHICLE LAW ISSUES OUTSIDE THE AEVA 2018

CHAPTER SEVEN
THE NEW REGULATORS
EMMA NORTHEY

INTRODUCTION

The system of regulation of road vehicles and their drivers comprises several elements, including: approval of vehicle design; licensing of drivers; registration of individual vehicles; confirmation of the ongoing roadworthiness of vehicles; monitoring (and, where appropriate, sanctioning) driver behaviour; and accident investigation.

One of the principal anticipated benefits of the introduction of CAVs is a significant reduction in the number of deaths and injuries resulting from collisions as a result of the removal of 'the human factor' from the driving equation.

In order to ensure the safety of all road users as CAVs enter service on the public highway, the regulatory environment will have to adapt, firstly to an environment in which driver assistance systems become more and more complex (up to and including SAE Level 3), and eventually to an environment in which there are no longer any drivers at all. Some elements of the current regulatory regime should adapt easily to these developments, but in other areas quite significant changes may be needed. This chapter explores what the future regulatory environment might look like.

OBJECTIVES OF THIS CHAPTER

The objectives of this chapter are to:

- outline the current system of regulation of road vehicles and their drivers;

- consider the extent to which the current system can be used to regulate CAVs;

- explore what additional or alternative regulatory measures might be required for road vehicles and/or their drivers (users) as we move towards the deployment of Level 5 CAVs.

CURRENT REGULATION OF ROAD VEHICLES AND THEIR DRIVERS

In order to promote road safety generally, both vehicles and their drivers are currently regulated by a range of agencies.

Vehicle Certification Agency

First on the scene is the Vehicle Certification Agency (VCA)[1], which is responsible for the type approval of new road vehicles before they are placed on the market. The manufacturer submits several production samples of the vehicle to the VCA, which then tests them for their conformity with all relevant performance standards (such as those relating to noise and emissions, braking efficiency, and airbag deployment).

Driver and Vehicle Standards Agency

Meanwhile, the Driver and Vehicle Standards Agency (DVSA)[2] ensures that the future driver of any such vehicle is competent by approving people to become driving instructors and administering the theory and practical driving test processes.

Once a vehicle is on the road, the DVSA ensures its ongoing safety by approving people to be MOT testers and maintaining a database of test results, which confirms that vehicles have been regularly checked and

---

1 https://www.vehicle-certification-agency.gov.uk/index.asp

2 https://www.gov.uk/government/organisations/driver-and-vehicle-standards-agency

found to meet basic standards of roadworthiness. The DVSA also monitors recalls of vehicles, parts and accessories to ensure that manufacturers address any identified problems fully and promptly.

Sales of unroadworthy and/or unsafe vehicles (as the text on the reverse of an MOT test certificate makes clear, it is possible for a vehicle to be unroadworthy but not unsafe, and vice versa) may be dealt with by local authority Trading Standards services or the Police, under Section 75 of the Road Safety Act 1988[3] and Regulations 5 and 8 of the General Product Safety Regulations 2005[4].

Driver and Vehicle Licensing Agency

The licence status of drivers (whether learners, qualified or disqualified) and any endorsements on their licences (for exceeding the speed limit, using a mobile phone while driving, and similar misbehaviour) is recorded by the Driver and Vehicle Licensing Agency (DVLA)[5] in its database.

The DVLA also maintains a database of individual vehicles, including the registration mark that is currently associated with each one (linked to its unique Vehicle Identification Number). The database also contains information about the accident history of some vehicles via their insurance write-off status (formerly Category A, B, C or D and now Categories A, B, S and N), whether they have been imported into the UK, and the transfer of personalized and other cherished registration marks.

---

3 Section 75 makes it a criminal offence to supply a motor vehicle in an unroadworthy condition

4 Regulation 5 makes it an offence for a producer to place a product on the market that is not safe and Regulation 8 makes it an offence for a distributor to supply a product if he knows or should have presumed, on the basis of the information in his possession and as a professional, it is a dangerous product

5 https://www.gov.uk/government/organisations/driver-and-vehicle-licensing-agency

Accident Investigation

Road accidents that result in fatalities are investigated by police collision investigators and coroners. There may also be a full investigation where an accident has had serious consequences (e.g. life-changing, but not fatal, injuries) or there is some other feature of the incident that suggests further investigation would be in the public interest. Incidents that cause relatively minor injuries are not typically investigated further because the police simply do not have the resources.

Lessons that prevent or reduce the likelihood of similar future incidents may be learned during the inquest process. Many of the Reports to Prevent Future Deaths made by Coroners[6] relate to concerns that have been identified in such cases, for example where the road design, layout and/or signage has been implicated in the causation of the incident.

CURRENT REGULATION OF OTHER TRANSPORT SECTORS

Other modes of transport, such as rail, air and sea, also have regulators for the design, production and use of the vehicles/craft themselves and the qualifications of their operators, but there is a key difference when it comes to accident investigation.

Each of these three other modes has a dedicated accident investigation branch: the Marine Accidents Investigation Branch (MAIB)[7] investigates marine accidents involving UK vessels worldwide and all vessels in UK territorial waters; the Rail Accident Investigation Branch (RAIB)[8] investigates accidents on the UK main line networks, London Underground and other metro systems, tramways, heritage railways and the UK part of the Channel Tunnel; and the Air Accidents Investig-

---

6 Under Regulation 28 of the Coroners (Investigations) Regulations 2013

7 https://www.gov.uk/government/organisations/marine-accident-investigation-branch

8 https://www.gov.uk/government/organisations/rail-accident-investigation-branch

ation Branch (AAIB)[9] investigates civil aircraft accidents and serious incidents within the UK, its overseas territories and crown dependencies.

The role of each of these three bodies is to investigate the causes of accidents, identify those factors that could lead to a similar incident in the future, and thereby to learn lessons that might help to prevent further avoidable accidents from occurring. They are not prosecuting bodies and they are not permitted to apportion blame or liability.

Although the regulations under which each investigator operates (the Merchant Shipping (Accident Reporting and Investigation) Regulations 2012, the Civil Aviation (Investigation of Air Accidents and Incidents) Regulations 2018, and the Railways (Accident Investigation and Reporting) Regulations 2005) do not exactly mirror each other, they do deal in similar ways with disclosure of material obtained by these bodies during their investigations.

The regulations provide that, while the final report of each investigation is a public document, the records of the investigation are not. In essence, these investigators will not disclose information provided to or obtained by them during an investigation, unless the person whose information it is consents or a court orders disclosure. The objective is to encourage any person who may have information relevant to identifying the cause of accidents to give full disclosure to the investigating body.

The test for a court to order disclosure is also cast in slightly different terms across the three sets of regulations, but the underlying theme is that the court must balance two competing interests: the interest in justice being served in the individual case, and the interest in the investigating body being able to obtain all the information that it needs to determine the cause of the accident. The concern is that, if people are fearful of implicating themselves or others in future legal proceedings,

---

9 https://www.gov.uk/government/organisations/air-accidents-investigation-branch

they may not cooperate fully with the investigating body and the opportunity to prevent another similar incident may thereby be lost.

At the time of writing, one of the most recent decisions in this area related to the Shoreham air disaster and, in particular, to the cockpit footage taken during the final flight of the Hawker Hunter on 22 August 2015. In determining an application by the BBC and the Press Association for disclosure to them of that footage, Mr Justice Edis observed that,

> "The inhibition on disclosure of material which comes into the possession of the AAIB in the course of a safety investigation, or which is created by the AAIB in such an investigation is a matter of international importance. The 2018 Regulations, and EU Regulation 996 and Annex 13 which underlie them, are designed to enable people involved in air accidents to co-operate with the AAIB investigation freely and without fear and to encourage them thereby to provide accurate information promptly and without obfuscation. This is to enable those investigations to reach accurate conclusions about the causes of air accidents and incidents so that air safety in the future is achieved. The culture of openness is an essential part of a safe system of air travel. It is called "the just culture" in some of the evidence and I shall use that expression in this judgment. It is given a very high priority in the law."[10]

On the facts of this specific case, Mr Justice Edis was not satisfied that the benefit of disclosure to the media outweighed the potential adverse impact on future safety investigations, and the application was therefore refused.

Although there is currently no Road Accident Investigation Branch, there have been calls in recent years for one to be introduced, including from Brake, the road safety charity[11], and the RAC Foundation[12]. In March 2017, the Parliamentary Advisory Council for Transport Safety

---

10 *British Broadcasting Corporation & Anor v The Secretary of State for Transport & Anor* [2019] EWHC 135 (QB)

(PACTS), sought (unsuccessfully) an amendment to the Vehicle Technology and Aviation Bill to enable the creation of such a body, stating,

> "*The time has come to set up a UK road collision investigation body. We have dedicated Accident Investigation Branches for air, rail and maritime but not for road accidents. The UK carries out some excellent collision investigation but it is fragmented and inconsistent. We need to learn from air and rail, harness the new technical opportunities, and bring together the efforts of researchers, Police, coroners, local authorities and others more effectively.*"[13]

IS THE CURRENT SYSTEM OF REGULATION APPROPRIATE FOR CAVs?

Testing Automated Vehicles on Public Roads

Before a manufacturer is ready to submit a production sample of a CAV for type approval with a view to making sales of that model to the general public, it will necessarily have carried out significant testing of that vehicle, both at private test tracks and on the public highway.

The testing of CAVs on public roads is qualitatively very different from the road-testing of standard motor vehicles. In essence, it is a test of the vehicle both as a piece of mechanical equipment (acceleration, braking, handling) and as a driver (hazard recognition and response, compliance with the Highway Code, traffic signs and road markings etc.). Arguably, therefore, it requires an approach more similar to the testing of a human driver, namely a human monitor of the CAV with the ability to

---

11 https://www.brake.org.uk/campaigns/flagship-campaigns/road-crash-investigation-branch

12 https://www.racfoundation.org/wp-content/uploads/2017/12/Towards_an_Accident_Investigation_Branch_for_Roads_Steve_Gooding_December_2017.pdf

13 https://www.theihe.org/pacts-calls-on-government-to-set-up-road-collision-investigation-branch/

take control of the vehicle until a certain minimum standard of safety has been established.

It is notable that the first CAV-related pedestrian fatality in the United States resulted from a collision with an Uber test vehicle. Although there was a human safety driver in the vehicle at the time, it appears from the available video footage that she had become distracted and was therefore more of a passive passenger than an active co-driver when the incident occurred.

In order to ensure that any testing that takes place in the UK is safe, a new *Code of Practice: Automated Vehicle Trialling*[14] was published in February 2019 by the Centre for Connected and Autonomous Vehicles (CCAV), a joint Department for Business, Energy & Industrial Strategy and Department for Transport policy team. The aim of the CCAV is to make the UK a premier development location for CAVs by working closely with industry, academia and regulators[15].

The key principles of the Code of Practice are that trialling organisations must ensure that they have:

- a suitably licensed and trained safety driver or operator, in or out of the vehicle, who should supervise the vehicle at all times and should be ready, able, and willing to resume control of the vehicle;

- a roadworthy vehicle; and

- appropriate insurance in place.

The Code of Practice also states that those planning CAV tests should speak with the highway and enforcement authorities, develop

---

14 https://assets.publishing.service.gov.uk/government/uploads/system/uploads/attachment_data/file/776511/code-of-practice-automated-vehicle-trialling.pdf

15 https://www.gov.uk/government/organisations/centre-for-connected-and-autonomous-vehicles/about

engagement plans, and have data recorders fitted. In the event of an incident, the data from the recorder should be made available to the relevant authorities to enable them to analyse the circumstances leading to the event.

Although this is a non-regulatory Code of Practice and a breach will not therefore be proof of negligence, as with any such code, it is likely to be highly persuasive evidence of a breach of the tester's duty of care in any subsequent civil proceedings.

Vehicle Type Approval

Once the testing phase has been completed and a developer is ready to place a new CAV on the market, it will be necessary for some form of type approval process to take place.

In its 2018 Preliminary Consultation, the Law Commission dealt with the regulation of vehicle standards at this stage in Chapter 4[16]. Interestingly, the Commission did not ask any questions about the fitness-for-purpose for CAVs of the current type approval processes relating to the mass production of new vehicles. Instead, they focused on the potential need for a new safety assurance scheme to deal with a single perceived inadequacy, namely the authorising of systems to be installed as modifications to registered vehicles or in vehicles manufactured in limited numbers ("small series").

It appears that this approach generated confusion in some respondents to the Consultation, producing answers to the questions about "*a new safety assurance scheme*" that treated this as a proposal to replace the entire existing type approval processes with something new. Helpfully, this revealed that there is a broad consensus in favour of third-party testing of CAVs rather than relying on self-certification alone, and a

---

[16] https://s3-eu-west-2.amazonaws.com/lawcom-prod-storage-11jsxou24uy7q/uploads/2018/11/6.5066_LC_AV-Consultation-Paper-5-November_061118_WEB-1.pdf

general agreement that the VCA would be an appropriate body to undertake all forms of type approval of CAVs[17].

On 4 September 2019, a new safety regime, called CAV PASS[18], was announced by the Future of Transport Minister. This scheme is being developed by government, the industry and academia to ensure the safety and security of autonomous vehicles. At the time of writing, few details are available and, in any event, its current focus appears to be on CAV trialling, so it is not yet clear whether the government intends to use the VCA for type approval or whether it would favour a new CAV-specific body.

Ensuring the Ongoing Safety of Vehicles

There are two strands to ensuring the safety of CAVs once they are being used on the public roads: checking the mechanical safety of the vehicles themselves (braking efficiency, tyre tread depth etc) and monitoring the ongoing safety of the automated driving function.

In terms of a CAV's mechanical safety, there would seem to be no particular reason why the current system of testing by DVSA-approved persons at 3-year intervals could not continue. However, it seems highly likely that such tests may be rendered irrelevant by the CAV's own systems. An automated vehicle must necessarily "know" much more about itself than a standard vehicle and it will be able to communicate this information to its owners. This should enable any developing mechanical issues to be identified and addressed very swiftly.

Turning to the automated driving system, a three-year interval between tests to ensure its safety intuitively feels absurdly long, and an entirely new approach is therefore likely to be necessary. As with mechanical performance, much of the safety-assurance is likely to come "bottom-

---

17 https://s3-eu-west-2.amazonaws.com/lawcom-prod-storage-11jsxou24uy7q/uploads/2019/06/Automated-Vehicles-Analysis-of-Responses.pdf Chapter 4

18 https://www.gov.uk/government/news/new-system-to-ensure-safety-of-self-driving-vehicles-ahead-of-their-sale

up" from what the CAV learns through its own diagnostic checks of its hardware, firmware and software. However, there will also need to be "top-down" deployment of safety-critical updates to the CAV's software (e.g. to deal with viruses and potential vulnerabilities) and of other upgrades to the system.

In Chapter 4 of its 2018 Preliminary Consultation, the Law Commission proposed that there would be an "automated driving system entity" (ADSE) identified for each automated driving system. The ADSE would be responsible both for seeking the initial type approval and for ensuring the ongoing safety of the CAVs using that system. It would therefore be the logical entity to take responsibility for monitoring the effectiveness of the CAVs own diagnostic processes and for deploying upgrades and updates.

It has already been mooted that the list of CAVs that the Secretary of State is required to keep by Section 1 AEVA 2018 could relate to individual vehicles rather than to CAV makes and models. If that proposal is adopted, it may be possible to go a step further and use DVLA's database to show, not only each vehicle's Section 1 AEVA 2018 status but also its update/upgrade status, either by enabling each CAV to transmit this information directly to the DVLA or by channelling it to them through the relevant ADSE. It would then be possible for either the DVSA or the DVLA to use that information to carry out virtual auditing of the work of the ADSEs to ensure that they were discharging their safety responsibilities.

Driver Training and Licensing

In the last decade, there has been a significant increase in the number of driver assistance features that are being added to new vehicles. The result is that there are now many vehicles on the roads that meet SAE Level 1 and an increasing number that have attained Level 2. While this has been achieved without the introduction of any new training or licensing requirements for drivers of such vehicles, there are some signs that the current regime is inadequate.

For example, there have been numerous reports of drivers, whose vehicles are only at Level 2, behaving as though their vehicle was fully driverless e.g. applying make-up or even taking a nap at the wheel. It is presently unclear whether these individuals had, in fact, misunderstood the capabilities of their Level 2 vehicles, but there does appear to be an existing case for better driver education.

Matters are likely to be complicated further when Level 3 and 4 vehicles become available. The transfer of control between an autopilot and a human pilot has long caused difficulties in the aviation sector. Those issues may be compounded by the relatively short distances between motor vehicles on the public roads and the lack of a third dimension in which to manoeuvre out of an emergency. In short, a motorist has very much less time to react and fewer options than a pilot (he can accelerate or decelerate and swerve to the left or right, but he cannot soar above or dive below).

A decision will have to be made as to whether, as we progress towards Level 5 CAVs (at which point there will no longer be a driver as such to be trained or licensed), drivers will be able to adapt to Level 3 technology (perhaps with the help of better education through public information campaigns or additions to the theory test) or whether additional compulsory practical training and/or testing will be required.

An alternative approach would be to accept that the most challenging aspect of Level 3, namely the emergency hand back of control from the vehicle to the human driver, is something that humans cannot be trained to cope with. As we explored in Chapter 1, one way to respond to that conclusion would be to bypass Level 3 entirely on the basis that it is simply too dangerous; another would be to accept that Level 3 emergency handovers will result in accidents but to create a legal presumption that any adverse consequences of such a handover are the vehicle's fault (the concept described by the Law Commission as "blame time"). The human driver would be given the opportunity to take control and attempt to avert disaster but would not be negligent if he failed to achieve something that was always likely to be beyond him. Such an approach, which obviates the normal process of scrutinising the acts and omissions of the human driver, would only be likely to be acceptable to other road users if it could be shown that emergency hand backs of control would occur only extremely rarely.

Once Level 4 is reached, the handover process will occur only in limited and clearly defined situations, which should remove much of the danger. At the same time, the role of the human in the driving seat will change in that, for large parts of any journey, the vehicle will have complete control of the driving process and no human monitoring will be required. As noted already, a new term "user-in-charge" has been proposed to describe this role. It seems likely that additional education would be required, perhaps within the theory test, to help people understand the different responsibilities of a "driver" and a "user-in-charge", but it seems unlikely that any additional practical training would be needed.

For as long as there remains a mixture of CAVs and standard vehicles on the road, there is likely to be a need to train drivers and riders of non-CAVs how to interact with CAVs on a day-to-day basis. Common issues that human drivers sort out amongst themselves through gesture and eye contact, such as who has precedence when three vehicles arrive at a mini-roundabout simultaneously, may need to be codified and form part of the training of every future driver.

Investigation of Road Accidents

It seems almost inevitable that there will be some accidents during CAV testing phases and quite probable that incidents will continue to occur occasionally when CAVs are in daily use. Zero collisions should remain the goal, but it could be many decades before it is achieved.

There will be two purposes to the investigation of any road accident involving one or more CAVs. The first is the same as the principal purpose of current investigations: to find out what happened to the individuals involved and why. The second objective arises from the fact that CAVs are new technology, which is interconnected in a way that standard vehicles have not been. There will be a system to be investigated, in addition to the behaviour of the individual driver and his vehicle, and it is inherently more likely that lessons of wider application could be learned as a result.

Two separate issues arise in connection with these two purposes:

- should there be a Road Incident Investigation Branch[19] to investigate causes of incidents and learn lessons without apportioning blame?

- will those who are involved in assessing blame, in particular police investigators in the case of criminal liability, have the necessary expertise and information to discharge their functions?

Rather unfortunately, both the above matters were dealt with by a single question within the Law Commission's 2018 Consultation, which did appear to confuse a number of respondents. In particular, there was a clear concern that a new Road Incident Investigation Branch might be intended to replace police investigators, when this was clearly not the case.

Instead, it seems probable that the police will continue to investigate whether any criminal offences have been committed. However, they are likely to require additional resources both to train their own staff and to enable them to draw on the support of external experts. They may also have to work more closely with the VCA, who will have a full understanding of how the CAV should have performed in the light of the type approval process.

---

19 While the term "Accident Investigation Branch" is used in relation to the other modes of transport, road safety campaigners tend to prefer the term "Collision" on the basis that there are no accidents (someone or something is always at fault). However, it appears to this author that a wider term would be preferable in relation to CAVs as incidents of concern to the public may not necessarily involve collisions – a malfunctioning CAV may well cause injury to its own occupants without ever colliding with anything else.

At present, individuals who have suffered losses as a result of a road accident may also have to do a certain amount of investigating in order to prove their civil claim. While the strong remedy in Section 2 AEVA 2018 should ensure that the process of obtaining financial compensation is relatively straightforward, with little by way of a technical investigation being required, it does not address an injured or bereaved person's need to know what went wrong and why. It seems possible that, unless there is a mechanism for providing such answers, individuals who have not received satisfaction from the criminal investigation or inquest process will seek to bring alternative civil proceedings or even private prosecutions to get to the truth.

Although the primary focus of a Road Incident Investigation Branch would be to determine the cause(s) of an incident in order to learn lessons that could prevent a recurrence, its reports could also provide a valuable source of answers for the individuals directly affected and would also play a significant role in increasing public confidence generally in CAV technology.

As matters stand, public attitudes to CAVs seem to be mixed. For example, a survey for the Department for Transport, published in October 2018, showed that respondents were more likely to name disadvantages than advantages in relation to CAVs, with safety issues identified on both sides of the equation[20]. It appears that everyone can understand why it would be safer if other people were not allowed to drive, but no one wants to give up their own control of their own vehicle.

It seems probable that thorough, independent investigations of any CAV-related incidents, together with a high degree of transparency as to the findings, will be key to improving public confidence in CAVs. It appears unlikely that the public will tolerate the withholding of information obtained during the course of Road Incident Investigation Branch enquiries, as happens in the case of investigations by the Accident

---

[20] https://assets.publishing.service.gov.uk/government/uploads/system/uploads/attachment_data/file/752240/transport-and-transport-technology-public-attitudes-tracker-wave-1-and-2-report.pdf

Investigation Branches for other modes. In the absence of international standards, of the kind that govern the way in which the AAIB approaches its investigations, there is no reason why a Road Incident Investigation Branch should be subject to this restriction. Further, it is unlikely that it will be necessary for such a provision in order to encourage humans with information to disclose it to the investigators, when, by virtue of the nature of the CAV, all the key information is likely to be held electronically.

SUMMARY OF POINTS ON THE NEW REGULATORS

Vehicles and their drivers are currently regulated by a range of agencies
The current systems of type approval, vehicle registration and driver training and licensing may be relatively straightforward to adapt to CAVs
New systems are required for the regulation of CAV trialling, ensuring the ongoing safety of automated driving systems, and CAV incident investigation

CHAPTER EIGHT
UNINSURED VEHICLES
EMMA NORTHEY

INTRODUCTION

As noted in Chapter 2, the "strong" remedy in AEVA 2018 for a person who suffers damage as a result of an accident caused by an automated vehicle when driving itself is only available if that vehicle is insured. The position when an uninsured CAV causes an accident is presently unclear. In the case of an uninsured conventional vehicle, the claim will usually be dealt with by the Motor Insurers' Bureau (MIB), but, at the time of writing, the MIB has yet to state its position in relation to accidents involving uninsured CAVs.

OBJECTIVES OF THIS CHAPTER

The objectives of this chapter are to:

- outline the current system for compensating people who sustain personal injury or other damage as a result of a road accident caused by an uninsured vehicle;

- consider whether this system would be applicable and/or fit for purpose where an accident is caused by an uninsured CAV;

- explore options for addressing any deficiencies in the current arrangements.

THE CURRENT SYSTEM – THE MIB

Part VI of the Road Traffic Act 1988 (RTA 1988) deals with compulsory insurance against third party risks. Section 143(1) prohibits a person from using on the road (or other public place) any motor vehicle that does not have a policy of insurance in force in respect of third-party risks arising from its use by that person. Contravening this prohibition is a criminal offence (Section 143(2)).

Unfortunately, not every road user complies with the law and there is a significant number of uninsured vehicles on the UK roads. While the lack of insurance cover does not extinguish an injured party's cause of action in tort against the uninsured driver, the likelihood of that person having the funds available to meet any claim is low.

In order to ensure that the victims of uninsured drivers are not left without compensation, the Secretary of State for Transport has entered into a contractual agreement with the MIB (a not-for-profit organization funded by the insurance industry) to ensure that their claims are paid.

UNINSURED DRIVERS AGREEMENT 2015

The Uninsured Drivers Agreement 2015 (the Uninsured Agreement) sets out the circumstances in which the MIB is obliged to satisfy a claim from a person who has sustained losses (including personal injury) as a result of a road accident caused by an uninsured driver. The key to obtaining payment under the MIB scheme is clause 3(1), which provides as follows:

> "*Subject to the exceptions, limitation and preconditions set out in this Agreement, if a claimant has obtained an unsatisfied judgment against any person in a Court in Great Britain then MIB will pay*

the relevant sum to the claimant or will cause the same to be so paid."[1]

According to clause 1(4) an *"unsatisfied judgment"* means a judgment or order in respect of a relevant liability which has not been satisfied in full within seven days from the date upon which the claimant became entitled to enforce it, and a *"relevant liability"* means a liability in respect of which a contract of insurance must be in force to comply with Part VI of the RTA 1988.

Reading the relevant provisions of the Uninsured Agreement and Section 143 RTA 1988 together, it becomes clear that the following criteria must be satisfied in order for an application to the MIB to succeed:

- the claimant has obtained an unsatisfied judgment against a person, Q;

- Q was required to be insured against third party risks arising from his use of the vehicle on the road.

In essence, therefore, the claimant proceeds against an uninsured driver in exactly the same manner as he would against an insured driver but, if the judgment he obtains is not paid by the defendant, then (subject to the claimant having complied with the various obligations set out in the remainder of the Uninsured Agreement) the MIB steps in and ensures that payment is made.

What the Uninsured Agreement does not do is give a claimant any direct right of action against the MIB in cases where the person at fault was not insured. There is nothing akin to regulation 3 of the European Communities (Rights against Insurers) Regulations 2002, which makes an insurer directly liable to an injured person and permits proceedings to be issued against the insurer itself, whether or not the insured person is also pursued.

---

1 https://www.mib.org.uk/media/166917/2015-uninsured-drivers-agreement-england-scotland-wales.pdf

WILL THE EXISTING UNINSURED AGREEMENT COVER CAV-RELATED ACCIDENTS?

Vehicles that meet SAE Levels 1 and 2 are already in use on the UK roads. Since they are under the control of a human driver, who can be sued in tort if he has been negligent, any accidents can be dealt with under the existing Uninsured Agreement. The same will be true of Level 3 vehicles, in the event that they are introduced.

Once Level 4 becomes a reality, there will be two different scenarios to consider, according to whether the human driver is in control or whether it is in the automated mode. Where a human driver of a Level 4 vehicle causes an accident, a claim in negligence can be brought against him. If he was uninsured and the judgment therefore remains unsatisfied, the MIB will step in.

However, if a Level 4 vehicle in automated mode (or, in due course, a Level 5 vehicle) causes an accident, the claimant will not have access to the "strong" remedy provided by Section 2(1) AEVA 2018 unless that vehicle was insured.

If the CAV was not insured because it was not required to be (where the exemption for public bodies applies or the vehicle is in the public service of the Crown), then Section 2(2) AEVA 2018 applies. It provides that the owner will be liable for any damage. Anyone who suffers loss as a result of an incident caused by such a vehicle will therefore have a version of the "strong" remedy provided by Section 2(1) AEVA 2018.

Section 2(2) is of no assistance, however, to anyone who is injured, or suffers other damage, in an incident caused by a CAV that is deliberately or inadvertently uninsured. In fact, there is nothing at all within AEVA 2018 to assist such a person.

A claimant who finds himself in this position will be forced to bring exactly the sort of technically complex, multi-party, and potentially disproportionately expensive 'traditional' claim that (as discussed in

Chapter 1) the AEVA 2018 was designed to avoid. He will thus be in a much weaker position than a person who suffers loss in an incident caused by an insured CAV, and, unless his claim has a high value, he may struggle to find representation.

Further, if any judgment he does obtain is not satisfied, perhaps, for example, because the party ultimately at fault for the defect in the automated driving system is a small player and has not paid its own insurance premiums, there will be no trigger for the MIB to step in. The party at fault will not have been the person using the vehicle on the road at the relevant time and will not therefore have been required to be insured against third party risks arising from the use of the vehicle on the road. The second of the two criteria necessary for the MIB's obligations to arise will not have been met.

Thus, the combined effect of the RTA 1988, the AEVA 2018 and the Uninsured Agreement is that the MIB's potential liability to meet claims is coterminous with that of human drivers and their insurers. CAVs are simply not covered.

WHAT MIGHT BE DONE TO ENSURE THAT VICTIMS OF UNINSURED CAV-USERS ARE NOT DISADVANTAGED?

In order to put the victims of incidents caused by uninsured CAVs on an equal footing with those caused by insured vehicles it may be to extend to such persons the "strong" remedy in AEVA 2018.

Although the MIB is not a creature of statute, it already has a statutory definition as a result of its appearance in Section 95 RTA 1988, which relates to the notification of refusal of insurance on grounds of health, where it is used to define "authorised insurer".

One straightforward means of extending the reach of the "strong" remedy would be to insert a new 2A into the AEVA 2018 to make the MIB liable in the same way that an insurer is liable e.g.

"Where—

(a) an accident is caused by an automated vehicle when driving itself on a road or other public place in Great Britain,

(b) section 143 of the Road Traffic Act 1988 (users of motor vehicles to be insured or secured against third-party risks) applies to the vehicle at the time of the accident,

(c) the vehicle is not insured at that time, and

(d) a person suffers damage as a result of the accident,

the Motor Insurers Bureau (a company limited by guarantee and incorporated under the Companies Act 1929 on 14th June 1946) is liable for that damage."

The MIB could also be given the same rights as individual insurers under Section 5 AEVA 2018 to then bring its own claim against the person actually responsible for the accident.

There would then be no need to amend the Uninsured Agreement, which could continue to regulate the relationship between the MIB and claimants injured by human drivers.

UNTRACED DRIVERS AGREEMENT

The Untraced Drivers Agreement 2017[2] provides a route by which the victims of 'hit and run' accidents can claim compensation for damage or personal injury that has been caused by a driver who cannot be traced.

The key to obtaining payment under the Untraced Drivers Agreement is clause 3(1)(c), which provides as follows:

---

2 https://www.mib.org.uk/media/353664/2017-untraced-drivers-agreement-england-scotland-and-wales.pdf

"… the person who is alleged to be liable in respect of the death of or bodily injury to any person or damage to property is an unidentified person (and where more than one person is alleged to be so liable, all such persons are unidentified persons)…"

It is not unusual for the make and model of the vehicle at fault to be known, but the driver to be untraceable because its registration mark has not been obtained. One can foresee a circumstance in which all vehicles of a certain make and model are known to be Level 4 CAVs. If a human was driving, he may not be able to be traced, but, if the vehicle was in automated mode, the persons responsible for the incident would effectively be known.

The requirement that <u>all persons</u> who are alleged to be liable must be unidentified could stand in the way of a claimant succeeding in his application to the MIB, if the MIB require the claimant to prove that it was not the automated driving system, which has identifiable persons standing behind it, which caused the accident. The claimant is unlikely to have any way of knowing which mode the CAV was in at the time.

The simplest solution may be to amend the Untraced Drivers Agreement to deem any Level 3 and 4 CAVs to have been being driven by a human at the time when the accident occurred, so that it is only that individual who must be unidentified for a payment to be made.

SUMMARY OF POINTS ON UNINSURED VEHICLES

The "strong" remedy in AEVA 2018 for a person who suffers damage as a result of an accident caused by an automated vehicle when driving itself is only available if that vehicle is insured.
The MIB's Uninsured Drivers Agreement only assists where a claimant has been injured by a human driver.
The AEVA 2018 is of no assistance to anyone who is injured, or suffers other damage, in an incident caused by an uninsured CAV.
The AEVA 2018 could be amended to extend the "strong" remedy.
The Untraced Drivers Agreement may need to be amended to ensure that persons injured by Level 4 CAVs are not at a disadvantage.

CHAPTER NINE
PRODUCT LIABILITY CLAIMS
SCARLETT MILLIGAN

OBJECTIVES OF THIS CHAPTER

The objectives of this chapter are to:

- Provide a brief oversight of the existing causes of action in product liability;

- Explore the difficulties (and liability loopholes) that this area of law poses for CAVs;

- Discuss how the AEVA may (or may not) change this landscape;

- Highlight the possible avenues for future legal reform in this area.

PRODUCT LIABILITY CLAIMS TODAY

Although often referred to as an area of law in its own right, what we know as 'product liability' actually spans numerous areas of law, including tort law, contract law, and EU law. In this chapter we will focus on tort and contract law, and their likely application to CAVs.

The Consumer Protection Act 1987 ("the CPA 1987")

References to product liability claims are often synonymous with claims brought under the CPA 1987, despite the ability to also bring claims in contract and tort, as we discuss below. The CPA 1987 was brought into fruition by virtue of a European Council Directive[1], and was hailed by many as a codification – and emboldening of – product liability law for consumers.

The CPA 1987 appears, at least at first blush, broad in scope. Sections 2(1) and 2(2) provide (emphasis added):

*"(1)… **where any damage is caused wholly or partly** by a defect in a product, every person to whom subsection (2) below applies **shall be liable for the damage**.*

(2) This subsection applies to—

(a) the producer of the product;

(b) any person who, by putting his name on the product or using a trade mark or other distinguishing mark in relation to the product, has held himself out to be the producer of the product;

(c) any person who has imported the product into a member State from a place outside the member States in order, in the course of any business of his, to supply it to another."

Section 2(3) further provides that suppliers – as opposed to manufacturers and those set out in s.2(2) above – shall be liable if:

"(a) the person who suffered the damage requests the supplier to identify one or more of the persons (whether still in existence or not) to whom subsection (2) above applies in relation to the product;

(b) that request is made within a reasonable period after the damage occurs and at a time when it is not reasonably practicable for the person making the request to identify all those persons; and

---

1 Directive 85/374/EEC of 25 July 1985 on the approximation of the laws, regulations and administrative provisions of the Member States concerning liability for defective products

(c) the supplier fails, within a reasonable period after receiving the request, either to comply with the request or to identify the person who supplied the product to him."

As a result, if a defective product causes damage, claimants are capable of suing a range of individuals and organisations, whose liability will be strict (i.e. a claimant does not need to demonstrate fault). While this may seem straightforward, there are five main hurdles for claimants wishing to bring product liability actions.

Firstly, there are limitations on who may bring proceedings under the CPA 1987. Section 5 of the CPA 1987 specifies that recoverable damage *"…means death or personal injury or any loss of or damage to any property (including land)"*. Further, if one is claiming for property damage, there are additional restrictions: damages cannot be recovered for losses under £275[2], and Section 5(3) tells us that the property must be:

"(a) of a description of property ordinarily intended for private use, occupation or consumption; and

(b) intended by the person suffering the loss or damage mainly for his own private use, occupation or consumption."

The intention behind this section is to restrict the protections of the CPA 1987 to consumers, rather than extending them to commercial purchasers and property. In the context of CAVs, this division appears somewhat arbitrary, as can be demonstrated by a simple example: if a defective CAV were to veer off-course and collide with someone's garden gate, the resident could use the CPA 1987 to claim for the damage to his or her gate. If, instead, the defective CAV veered off course 300 metres down the road and collided into a glass shopfront, the owner of the shop could not bring proceedings under the CPA 1987, as the damage is to commercial property, rather than private

---

2 Section 5(4) CPA 1987

property. Instead, the shop owner would need to bring proceedings in negligence (which we discuss in more detail later in this chapter).

Secondly, there are limitations on which 'goods' may be recoverable under the CPA 1987. A likely source of frustration for owners of defective CAVs will be Section 5(2), which makes clear that:

> "*A person shall not be liable* under section 2 above in respect of any defect in a product *for the loss of or any damage to the product itself* or for the loss of or any damage to the whole or any part of any product which has been supplied with the product in question comprised in it." (emphasis added)

Thus, the owner of a defective CAV could not rely on the CPA 1987 to recover for the damage to the defective CAV, even if it had been rendered completely unroadworthy due to a collision. Again, he or she would need to bring a separate (although perhaps concurrent) claim in negligence (and/or a contract claim, which we also discuss later in this chapter).

As at the time of writing, cars with elements of CAV technology are considerably more expensive than Level 0 'ordinary' vehicles. Given that CAV technology will continue to develop, and manufacturers will continue to incur significant research and development costs, it is unlikely that there will be a noticeable reduction in the price of CAVs for quite some time. Consumers will understandably be concerned that they are unable to recover considerable sums of money through their main avenue for product liability protection.

Thirdly, there is a need to establish which 'product' is allegedly defective. CAVs consist of numerous components of hardware and software, which may have been manufactured by a number of companies. This will, of course, affect whom one chooses to sue.

There is another pressing reason to identify the 'product' in question: if the CPA 1987 prohibits recovery for the defective product itself, it is important to identify whether the defect lies within the CAV itself, or if the problem can be categorised as a defect in a separate, standalone product.

The importance of this distinction can be illustrated with an example concerning defective tyres: if the tyres were supplied with the CAV, they would be deemed to be part of "*the product itself*", or perhaps to be tyres "*supplied with the product in question comprised in it*"[3]. As a result, the damage would be to the CAV ("*the product itself*"), and would not be recoverable under the CPA 1987. By contrast, if the owner had purchased some tyres separately to improve the appearance of his or her car, those tyres would constitute a separate 'product'. In that instance, whilst the owner could not recover damages for the defective tyres, he or she could claim for any resulting damage to the CAV, which would be property damage *caused by* the defective product.

There is an added layer of complexity when we consider the software components of CAVs. There are significant doubts as to whether software is deemed to be a 'product' falling within the ambit of the CPA 1987. Section 1(2) defines a product as: "*…any goods or electricity…*". The question is this: can software, an intangible object, be appropriately described as a 'good', a word with strong physical connotations?

We are unaware of any cases determining the point in the context of the CPA 1987; however, case law in the context of contractual legislation strongly suggests that the answer is 'no'.

---

3 ibid

One of the first landmark authorities[4] on this point is the obiter dicta of Lord Justice Glidewell in *St Albans City & District Council v International Computers Ltd*[5], who said: "*…it is necessary to distinguish between the program and the disc carrying the program*", and that whilst the disc containing the program would be classed as 'goods', "*the program itself is not "goods" within the statutory definition*". On this reasoning, a product which incorporated both hardware and software (for example, a computer, or a CAV with in-built software) would meet the definition of 'goods', but software purchased without accompanying hardware would not. A number of cases have since followed this approach[6]. Whilst CAVs are likely to be pre-supplied with software, it is abundantly clear that their software will require updating, and may be sold separately in the case of 'add on' features or upgrades. Such software will, in all probability, be provided electronically (and perhaps through cloud-based software) as opposed to via a hard medium (through a USB drive, for example).

In the recent case of *Computer Associates UK Limited v Software Incubator Limited*[7], the Court of Appeal authoritatively ruled that, for the purposes of the Commercial Agents (Council Directive) Regulations 1993, computer software which was supplied by electronic download was not "goods". Whilst Lady Justice Gloster accepted that

---

4 Earlier cases not discussed in detail include: *Eurodynamics Systems Plc v General Automation Limited* (unreported, 6 September 1988), in which Steyn J (as he then was) did not wish to determine the point, as it was not necessary for the resolution of the case; and *Saphena Computing Limited v Allied Collection Agencies Limited* [1995] FSR 616, in which the Court of Appeal did not disturb the finding of the Recorder below that "*…it was an implied term of each contract for the supply of software that the software would be reasonably fit for purpose…*" (at page 644) . In the latter case, the Recorder did not explain how or why this position was reached. Thus, the authors consider these authorities to be less helpful than those set out in this chapter.

5 [1996] 4 All ER 481

6 See, for example, *Horace Holman Group Limited v Sherwood International Group Limited* (2002) 146 SJLB 25, where the High Court, for the purposes of making a determination under the Unfair Contract Terms Act 1977, found that computer software did not constitute "goods". See also footnote 8.

7 [2018] EWCA Civ 518

this approach was outdated and *"seems artificial"* in light of the technological developments that have taken place in recent years, she felt compelled to reach this conclusion given the wealth of authority on the point, including the aforementioned case of *St Albans City*.

This decision has been appealed to the Supreme Court. At the time of writing, the justices had heard the appeal and decided to refer the question of whether software could amount to "goods" to the Court of Justice of the European Union ("CJEU"), which should provide some welcome clarity on this issue.

If the Supreme Court (by way of the CJEU) were to reverse the Court of Appeal's decision, this could result in significant, sweeping changes across a number of legislative regimes, including the CPA 1987. A number of the authorities referred to within *The Software Incubator* proffer the view that such a substantial change ought to be made by Parliament, rather than the courts[8].

One factor which appears to fortify this view, as was recognised by Gloster LJ in *The Software Incubator*, is the decision by Parliament when enacting the Consumer Rights Act 2015 (discussed later in this chapter) to introduce specific contractual rights and obligations for the *"supply of digital content"*, separate from the obligations in respect of *"goods"*[9]. Section 2 of that act defines these terms as:

---

8 For a particular example, see the judgment of Lord Justice Moore-Bick in *Your Response Limited v Datateam Business Media Limited* [2014] EWCA Civ 281, at paragraph 27.

9 Interestingly, the reasoning behind the introduction of a separate definition of, and protections for, 'digital content' was *not* an affirmation that digital content was not protected by the definition of 'goods' or 'services', but because of the uncertainty surrounding this issue, which was deemed unsatisfactory for consumer rights. See the Department for Business, Innovation and Skills' July 2012 Consultation Paper entitled "Consultation on the supply of goods, services and digital content", particularly paragraphs 4.18 and 7.7: https://assets.publishing.service.gov.uk/government/uploads/system/uploads/attachment_data/file/31350/12-937-enhancing-consumer-consultation-supply-of-goods-services-digital.pdf

Of course, beyond consumer rights law this uncertainty remains and, ironically, by taking proactive steps to protect consumer rights, the legislature may have inadvert-

"(2)... (8) "Goods" means any tangible moveable items, but that includes water, gas and electricity if and only if they are put up for supply in a limited volume or set quantity.

(9) "Digital content" means data which are produced and supplied in digital form."

The main benefits of CAVs undoubtedly lie in the capabilities produced by their software and artificial intelligence, without which they would be largely unrecognisable from our current vehicles. An inability of the cornerstone piece of product liability legislation to combat defective software would be a major deficiency in our legal framework. As we explore in later sections of this chapter, bringing alternative (contractual and/or tortious) claims against manufacturers is not without difficulties and therefore cannot be seen as an adequate 'backup' remedy.

Fourthly, although the Act creates a regime of strict liability, this is by no means a home run for claimants. Whilst a claimant does not need to demonstrate fault on the part of manufacturers, he or she will still need to satisfy a court, on the balance of probabilities, that a product defect existed and that it was the whole or partial cause of the damage sustained.

Section 3(1) of the CPA 1987 defines 'defect' in the following way:

*"(1) Subject to the following provisions of this section, **there is a defect in a product for the purposes of this Part if the safety of the product is not such as persons generally are entitled to expect**; and for those purposes "safety", in relation to a product, shall include safety with respect to products comprised in that product and safety in the context of risks of damage to property, as well as in the context of risks of death or personal injury."* (emphasis added)

---

ently removed the legislative protections of others.

What are the general public "*entitled to expect*" when it comes to the safety of cars? They cannot reasonably expect cars to be safe havens which offer protection from all possible events and dangers. Accidents do happen, and are seen by most as an unfortunate side effect of the benefits that cars bring.

It is equally unreasonable to insist that our cars protect us from every externality, or that a car will never harm other road users or pedestrians. Such harms are, of course, to be avoided or minimised, but their occurrence alone does not render a car unsafe. By way of analogy, while a knife may be used in a way which threatens someone's safety, that alone does not render the knife - as a product - unsafe or defective for the purposes of product liability law.

Furthermore, the general public's expectations must be judged by reference to society's expectations and standards in place *at the time of a car's manufacture*, rather than by current expectations. To take a far-fetched example: suppose that CAV technology develops in such a way that a CAV can – with incredible precision and safety – detect when an injury-causing collision is about to occur, and eject its passengers to safety by detaching and ejecting its inner pod. If one manufacturer rolled out this technology in 2030, it would be unreasonable to suggest that all CAVs manufactured prior to that date were defective.

The question of what society is "*entitled to expect*" is especially difficult in the context of Level 3 CAVs. Unlike a Level 5 CAV, we are not entitled to expect these vehicles to drive us safely from Point A to Point B without malfunctioning. By their very definition, Level 3 CAVs may find themselves unable to deal with the task of driving, and will request the User in Charge to take over. We foresee the question of safety expectations being determined by the surrounding circumstances in which a Level 3 CAV abdicates responsibility for driving, including the extent to which the User in Charge is given advance warning (if any at all).

Even with Level 5 CAVs, society will not be "*entitled to expect*" that they can be pre-programmed to deal with every eventuality they encounter on the roads. As we set out in Chapter 1 ("Life Before the AEVA"), CAVs will necessarily be governed by artificial intelligence which will learn from a CAV's interactions and environment, enabling them to improve their decision-making over time. Whilst great efforts will no doubt be undertaken by manufacturers to make their CAVs as ready for the road as possible, most commentators are expecting a degree of trial and error when this technology first appears on our roads. Would a court therefore hold that we are not entitled to expect any CAV to be error free? Might a court even decide that errors made by a CAV are not a product 'defect', on the basis that the CAV was built and supplied as a self-learning robot, and has performed as such?

A CAV's artificial intelligence will also need to make split-second decisions about how to react to imminent dangers, and whose safety (or life) to prioritise[10]. As a society, what do we expect CAVs to do in those situations? Whose life should they prioritise, and are there any exceptions to that rule? Should they be pre-programmed to always act in a way which saves the greatest number of people? Or instead to always prioritise their occupants? These are interesting and difficult ethical debates, and there is unlikely to be one universal 'expectation' across society. This places a court in an extremely difficult position when faced with deciding whether the CAV was 'defective' for acting, or having been programmed, in a particular way.

When grappling with these difficult questions, the courts are instructed to take into account[11] a product's:

- marketing;

- get up;

- markings;

---

10 An ethical dilemma known as the 'trolley problem'

11 Section 3(2) CPA 1987

- instructions;

- warnings;

- reasonable and expected use; and

- the time of its supply.

Given the subjective difficulties discussed above, it is our view that clear and instructive marketing and/or communications with consumers will be a pivotal factor in a court's determination of an alleged defect. For example, Level 2 or Level 3 CAVs may be marketed as 'autonomous' or 'driverless' (and with minimal warnings as to the level of input required), or instead as a vehicle with 'driving assistance technology' (with clear warnings as to what Users in Charge can expect and be prepared to do). These statements will carry a great deal of weight in a court's assessment, as they directly inform what the general public will expect from that product.

These are all tricky questions which the courts will need to answer in the absence of any statutory intervention by Parliament. The difficulty of this exercise will be compounded by the information and evidence gathering required for litigation, which will be aimed at extracting what went wrong within the CAV. When one bears in mind that this may involve understanding how numerous elements of hardware and software interact, and that all of these elements could have been manufactured by different entities, the scale of this exercise becomes apparent.

It is clear to us that resolution of these issues will require expert evidence. However, experts in this field will, initially, be far and few between. In fact, the experts most qualified to speak to a CAV's performance (or that of its components) are likely to be those employed by (or consulted by) the CAV's manufacturer. Given the conflict of interest this will create, we foresee significant difficulties for claimants seeking to establish that a CAV was defective and, in stark contrast, we foresee defendants gaining a significant advantage in product liability litigation by virtue of their superior knowledge and access to expert advice.

Furthermore, manufacturers may be unwilling to disclose product information on the basis of commercial sensitivity. One possible resolution to this conundrum could be to introduce a rebuttable presumption to the effect that where CAV software played a role in a collision, it is deemed to have been the cause of the accident. It would then be for a manufacturer to make a decision as to whether it wished to disclose material and expert advice to rebut that presumption, and avoid liability. Clearly this is not a solution without problems, not least the possible injustice to manufacturers, and the possibility of creating barriers to CAV development in England and Wales.

Fifthly, the Act confers a myriad of possible defences on defendants. Many of these are problematic in the context of emerging technology.

> *"4 (1) In any civil proceedings by virtue of this Part against any person ("the person proceeded against") in respect of a defect in a product it shall be a defence for him to show—*
>
> > *(a) that the defect is attributable to compliance with any requirement imposed by or under any enactment or with any EU obligation; or*
> >
> > *(b) that the person proceeded against did not at any time supply the product to another; or*
> >
> > *(c) that the following conditions are satisfied, that is to say—*
> >
> > > *i. that the only supply of the product to another by the person proceeded against was otherwise than in the course of a business of that person's; and*

ii. that section 2(2) above does not apply to that person or applies to him by virtue only of things done otherwise than with a view to profit; or

(d) that the defect did not exist in the product at the relevant time; or

(e) that <u>the state of scientific and technical knowledge at the relevant time was not such that a producer of products of the same description as the product in question might be expected to have discovered the defect if it had existed in his products while they were under his control</u>; or

(f) that the defect—

i. constituted a defect in a product ("the subsequent product") in which the product in question had been comprised; and

ii. was wholly attributable to the design of the subsequent product or to compliance by the producer of the product in question with instructions given by the producer of the subsequent product."

(emphasis added)

Whilst all of these defences will be relevant to CAVs to differing degrees, we have highlighted Section 4(1)(e) (the so called 'development risks defence') as it seems particularly relevant to the ever-developing advancements in CAV technology. Currently, we are in the early throes of a long-term change to the manufacture, capability and use of vehicles. It is inevitable that manufacturers' understanding of CAV technology will grow exponentially, and will be informed by past mistakes. This poses a significant problem for early users of defective CAVs: a manufacturer could legitimately argue that, at the time of a particular CAV's manufacture, it was not aware of how the technology would perform, and thus that the defect was not "*known*".

The scope of this problem is enlarged when one considers the self-learning nature of CAVs. To some extent, manufacturers will be unable to control how and what these machines learn. Whilst they may be able to overwrite 'erroneous' or 'bad' learning, they are unlikely to be able to fully prevent it. The actions of CAVs may therefore be said to be outside of a manufacturer's knowledge at the time that the product was created, and that this would have been the case even if the manufacturer had undertaken further investigations or testing. Additionally, the language of this defence (namely that the manufacturer *"...might be expected to have discovered the defect if it had existed in his products while they were under his control"*), could also rule out any claims on the basis of 'bad learning' which occur *after* the CAV has left the manufacturers' control. That said, claimants may have a successful counter-argument on the basis of a manufacturer's leasing, scrutinising, and provision of software updates, which all offer ongoing opportunities to control the CAV.

If this point were decided in favour of CAV manufacturers, this would rule out liability for a significant proportion of CAV complaints, and in circumstances where society might expect manufacturers to bear the financial burden of injuries and damage caused by the initial roll-out of CAVs.

Claims in Contract

General issues

Contracts for the sale, hire, or use of vehicles are concluded up and down the country every day, and will contain a whole host of terms and conditions, whether standardised or bespoke. In either case, the terms and conditions will require interpretation: what was actually being promised, guaranteed or indemnified? Is the loss claimed for actually covered by the terms of the contract? Does the contract cover third party losses?

This book is not the appropriate forum to suggest precise contractual express terms which are likely to benefit drivers, insurers, manufacturers, or others. It would, however, be remiss of us not to flag the role of implied terms in the context of vehicle-related product liability claims. In addition to examining the express terms of a contract, a court may find that terms have been implied into the contract. Terms can be implied by the courts themselves[12], or by function of a statute.

Statutory implied terms

For our purposes, some of the most important statutory implied terms are those implied into contracts between buyers and sellers of vehicles. We are primarily concerned with the Consumer Rights Act 2015 (applicable to contracts with consumers only), or with the Sale of Goods Act 1979 and/or the Supply of Goods and Services Act 1982 for non-consumer contracts. These pieces of legislation may, in varying situations, imply terms that the vehicles will be (inter alia):

- of satisfactory quality, taking account of their description, price, and all other relevant circumstances;

- fit for the purpose for which they are commonly supplied, i.e. driving in the manner usually associated with the vehicle in question;

- safe;

- durable.

As with the CPA 1987, the rapid development of technology, and how CAV capabilities are communicated and marketed, will be pivotal factors in a court's assessment of whether a CAV satisfies the implied terms above.

Again, the standards to which the courts will hold buyers are likely to change with time, and may generate uncertainty initially: the safety and

---

12 Typically done so to: give effect to the intention of the parties; reflect commercial common sense; reflect business necessity; or to reflect custom and usage.

reliability of CAVs is not yet fully understood, and it is likely that the courts will apply increasingly robust standards with time and technological advancements.

Will faulty software amount to a breach of contract?

As we highlighted earlier in this chapter, the Consumer Rights Act 2015 has drawn a distinction between "*goods*" and the "*supply of digital content*", and dictates implied terms for both. As a result, and noting the outstanding ruling from the Supreme Court in *Computer Associates*, it seems likely that the courts would interpret the other statutes (the Sale of Goods Act 1979 and the Supply of Goods and Services Act 1982) as *not* applying to software.

If that were the case, a commercial purchaser (for example, a commercial car club owner who purchased specific software for analysis of car journeys) could not rely on the implied terms above when bringing a claim for defective software, and would need to rely on the actual terms of his or her contract with the seller (or bring an alternative cause of action).

Where the courts do have cause to assess the performance of CAV software, it is highly unlikely that they will expect it to perform perfectly. Not only is this wholly unrealistic, case law suggests that the courts might take a more lenient approach to defective software. For example, in *Saphena Computing Limited v Allied Collection Agencies Limited*[13], the Court of Appeal held:

> "…it is important to remember that software is not necessarily a commodity which is handed over or delivered once and for all at one time. It may well have to be tested and modified as necessary. It would not be a breach of contract at all to deliver software in the first instance with a defect in it".

---

13 See footnote 4, at page 652

In *St Alban's*, the Court of Appeal noted the findings of *Saphena* and somewhat tempered the effect of those remarks in holding that[14]:

> "Parties who respectively agree to supply and acquire a system recognising that it is still in course of development cannot be taken, merely by virtue of that recognition, to intend that the supplier shall be at liberty to supply software which cannot perform the function expected of it at the stage of the development at which it is supplied."

Both cases were concerned with contractual relationships for the development of *bespoke* software services, which may go some way in explaining the courts' willingness to understand that a period of feedback and dialogue may be required. In our view, courts are unlikely to be as forgiving when faced with software defects in finished products delivered en masse to the consumer market (i.e. mass-market CAVs).

Other limitations on the use of statutory implied terms

As we have discussed elsewhere in this book (particularly in Chapter 1), claims are likely to arise as a result of problematic handovers between the CAV and the User in Charge. As with CPA 1987 claims, it is perfectly foreseeable that the courts may find handovers to be inherent aspects of Level 3 and Level 4 CAVs, rather than a feature which renders a CAV in breach of the implied terms to be safe, fit for purpose and of satisfactory quality. To return to our knife analogy, knives as products do have an inherent danger, but provided they are of an appropriate quality and design, they will nonetheless be deemed safe, fit for purpose, and of satisfactory quality from the vantage point of these implied terms.

Another limitation to the efficacy of the aforementioned implied terms is the scope of individuals who will be able to rely on them: unless expressly provided for by the contract, only a person who has entered into the contract (i.e. a purchaser or a hirer of a CAV) will be able to rely on a contractual cause of action.

---

14 See footnote 5, at page 258

This will therefore exclude swathes of people who may wish to bring claims for injury and/or damage caused by CAVs, including: passengers; those owning or travelling in vehicles involved in the collision; and those who are external to the CAV but are somehow involved (e.g. pedestrians who are injured, or those who have their property damaged).

Claims in Negligence

Issues and exclusions

Although typically associated with road traffic accidents, negligence claims may also be brought in respect of defective products. Claims may be brought against parties such as manufacturers, suppliers, or repairers where their failure to take reasonable care in the manufacturing or repairing process ultimately causes personal injury and/or property damage[15]. Unlike actions under the CPA 1987, actions in negligence are not strict, and require a claimant to establish fault.

As the law currently stands, for some third parties[16] a negligence suit will be their only option to recover for damage or injury caused by a defective CAV: for example, if they do not have a contract with the defendant, and are claiming for commercial property damage[17], or if one of the CPA 1987 defences apply. Furthermore, in light of the uncertainty surrounding how the CPA 1987 will be applied to CAVs, it may be prudent for many claimants to bring a concurrent claim in negligence. As is likely to have become clear to readers by now, the *type* of product liability claim one brings will very much depend on who you are, the relationship you hold with the manufacturer/supplier/repairer (if any), and the type of loss you are seeking to recover.

---

15 See, for example, *Herschtal v Stewart & Ardern Limited* [1940] 1 K.B. 155, concerning the negligent reconditioning of a car, or fixing of its wheel, which came off whilst the vehicle was being driven

16 Which, for the avoidance of doubt, includes innocent bystanders: see, for example, the case of *Stennett v Hancock and Peters* [1939] 2 All ER 578

17 Which is not covered by the CPA 1987 - see the 'Consumer Protection Act 1987' section

PRODUCT LIABILITY CLAIMS • 137

One cannot recover damages in negligence for the defective property itself (e.g. a defective CAV)[18]. This loss is categorised by the courts as a 'pure economic loss' and will be irrecoverable in the absence of a 'special' *Hedley Byrne* relationship[19] (principally where the other party has a duty to safeguard a claimant against economic losses). The prohibition on the recovery of pure economic losses would also inhibit the recovery of losses stemming from damage to a CAV; for example, a car club's loss of profits caused by an inability to fulfil a hire contract.

This again throws up factual questions about what exactly the defective product is, and whether it can be classified as separate from the CAV. In negligence cases the courts have taken a similar approach to the CPA 1987: generally speaking, they have asked the following questions to determine whether a product can be classified as 'standalone':

- Was the product incorporated into a larger product at the time of purchase/acquisition? For example, were the tyres supplied with the CAV, or were they subsequently purchased and installed by the claimant?[20]; and

- Was the product an integral part or key component of the larger product? Sticking with the same example, the tyres on a CAV at the time of purchase are an 'integral part' of the vehicle: it cannot be driven away and used in its ordinary manner without them. By contrast, the subsequent purchase of tyres which are then fitted to the CAV are not an integral part of a CAV at the time of purchase[21].

---

18 See a key authority on this point, *Murphy v Brentwood District Council* [1991] 1 A.C. 398, particularly at page 475

19 Referencing the eponymous case of *Hedley Byrne & Co Limited v Heller & Partners Limited* [1964] A.C. 465

20 See *M/S Aswan Engineering Co v Lupdine Limited* [1987] 1 W.L.R. 1 at page 21, where this particular example of car tyres is discussed by Lord Justice Lloyd

21 See, for example, the cases of *Linklaters Business Services v Sir Robert McAlpine Limited* [2010] EWHC 2931 (TCC) at paragraphs 115-119, and *Bellefield Computer Services Limited v E Turner and Sons Limited* [2000] 1 WLUK 688,

Where some of the intricate components of CAVs are concerned (most likely to be software components) these questions could entail a great deal of argument and expert evidence.

Negligence in respect of defective vehicles

In advancing the claim that manufacturers or repairers have been negligent in the production or repair of a CAV, a claimant would, generally speaking, need to establish that the manufacturer or repairer in question had failed to take reasonable care in their manufacturing/repairing process. This is not the same as proving that the item was defective, which may not be in dispute. A claimant would, for example, need to demonstrate failings such as:

- poor design of the CAV;

- an inadequate system of manufacture or testing (for example, a failure to adequately screen, test, or control the products and processes, or to ensure that appropriate subcomponents are used); or

- other negligent failings by one of the manufacturer's employees, notwithstanding the systems in place.

Such failings may not be straightforward to prove, and the existence of a defect alone is unlikely to be sufficient to get a claimant home.

The difficulties in identifying a breach of the duty of care will be exacerbated where several defendants are involved in the manufacture of a CAV (for example suppliers of raw materials or raw software). In such instances, it may be difficult for a claimant to identify where the defect lies. Unless this can be resolved with pre-action correspondence or disclosure, claimants may need to sue any number of defendants protectively. Ultimately, it will be for the claimant to *positively* prove, on the balance of probabilities, that at least one of the defendants was

---

particularly the judgment of Justice May

negligent; trying to demonstrate the negligence of one or more defendants through a process of elimination will not suffice.

This does not necessarily mean that a claimant must identify the *precise* negligent act within the manufacturing process. Provided that the claimant can demonstrate, on the balance of probabilities, that the defect was attributable to one of the defendant's processes (rather than some external cause), a court is, in principle, able to find that a manufacturer has been negligent. The courts have recognised that pinpointing an act with further precision may require knowledge only held by the manufacturer[22].

Furthermore, even where there are multiple suppliers and manufacturers, the end manufacturer/assembler of a CAV does have an ultimate responsibility to ensure that the components supplied have been chosen with reasonable care, and that they are inspected to ensure that:

> "…those parts can properly be used to put his product in a condition in which it can be safely used or consumed in the contemplated manner by the ultimate user or consumer."[23]

Thus, provided an end manufacturer has taken reasonable care in selecting components, ensuring they are compatible, and inspecting the

---

[22] Cases which demonstrate this principle in a vehicle context are: *Carroll and Others v Fearon and Others* [1999] E.C.C 73, concerning a tyre which burst many years after manufacture, where it was established that the tyre defects must have been attributable to the manufacturing system. The Court of Appeal held that a judge need not go further and identify particular individuals, acts or omissions in the manufacturing system: "*If the manufacturing process had worked as intended this defect should not have been present*". See in particular paragraphs 17-26. See also *Alan Peter Ide v ATB Sales Ltd; Lexus Financial Services t/a Toyota Financial Services (UK) PLC v Sandra Russell* [2008] EWCA Civ 424, in which the claimant successfully established that a fire was caused due to a defect in the electrics of her Lexus car. She was not required to explain, prove, or describe the specifics of the defects.

[23] Taken from Charlesworth on Negligence (4th ed.), para. 797, as cited with approval in *Taylor v Rover Company Limited and Others* [1966] 1 W.L.R. 1491, at page 1499

overall product, that manufacturer is unlikely to be held responsible for *all* irregular incidents or defects.

When faced with negligence actions in respect of the manufacture or repair of a defective CAV, the courts are likely to place considerable weight on:

- the manufacturer's/repairer's overall processes and safety checks;

- the manufacturer's/repairer's knowledge and understanding of the CAV (and its components) at the material time;

- the level of understanding across the CAV industry at the time;

- whether the particular CAV met governmental/industry standards at the time of manufacture/repair;

- whether the defect arose from a failure in a component, or as a result of poor decisions made by the CAV's artificial intelligence.

Tangible and behavioural failures (such as: ascertainable manufacturing defects; poor designs; or failures to test and produce CAVs to industry standards) are more likely to be successful than cases concerning the performance of the CAV's software and AI, as it will be significantly easier to identify the unreasonable behaviour/breach of duty in the former category of cases.

In addition to manufacturers and repairers, negligence claims may be brought against car dealers or second-hand sellers. Where sellers have reason to believe that a CAV could cause harm (either because they are cognisant of specific facts, or could have discovered them had they been reasonably diligent), they will be held negligent if they fail to disclose this to the purchaser. However, sellers will not be negligent where there is a reasonable anticipation that the buyer will examine the CAV before using it[24], though this is unlikely in the context of mass-market CAVs.

A seller is also likely to be found negligent in circumstances where it fails to make clear that safety checks have not been undertaken, or where it otherwise fails to give sufficient warnings in respect of known defects[25].

Negligent failures to warn or recall

Aside from defective products, negligence claims may also encompass a failure to warn, a failure to give any or any adequate safety instructions, and – arguably – a failure to recall defective products. This appears to be fertile ground for claims involving Level 3 CAVs and early users of CAV technology. Handovers to Users in Charge are, in and of themselves, unlikely to be seen as negligent defects, being an inherent part of the product and a known risk to the user. However, a CAV's failure to give clear and adequate warning of a handover, or a manufacturer's failure to give adequate instructions on handovers, may have more success before the courts, and could potentially give rise to claims against other distributors[26]. As highlighted earlier in this chapter, clear communications with purchasers will be key.

---

24 See *Andrews v Hopkinson* [1957] 1 Q.B. 229, in particular pages 236-237

25 For discussion on this point, see *Hurley v Dyke* [1979] R.T.R. 265, where liability was not established. It was found that the seller, at the highest, had knowledge that the vehicle *might* be defective, and that it required further examination. On that set of facts, the sale of the vehicle "*as seen and with all its faults and without warranty*" was sufficient to discharge the duty of care. See pages 301 and 302 in particular.

26 See case law from Canada on this point: *Rivtow Marine Limited v Washington Iron Works* [1974] S.C.R. 1189 (S.C.C.), and the subsequent case of *Hutton v General Motors of Canada Limited* [2011] 4 W.W.R. 284, both of which affirmed that the duty to warn purchasers may apply beyond manufacturers and to distributors in circumstances where those distributors: know the purpose for which a product is to be used; know the danger posed by the defect; and fail to warn those purchasers. The latter case of *Hutton* involved a failure by a car distributor to warn the purchaser of a defective sensor in their car, resulting in the airbags deploying and causing the claimant injuries.

Throughout this book, we have repeatedly emphasised that the courts will assess claims having regard to the dangers known to manufacturers and the CAV industry *at the material time*. This is also true in the 'failure to warn' category of cases, though there is one possible exception: having placed a potentially dangerous CAV on the market, manufacturers may be obliged to inform customers that this CAV is now known to be dangerous, and perhaps subject to a product recall.

Outside of negligence actions, the CPA 1987 confers a power on the Secretary of State to serve notices which prohibit manufacturers from supplying goods, or require them to publish a warning about unsafe goods[27]. Failure to comply with such a notice is a criminal offence[28].

In addition, the General Product Safety Regulations 2005 confer powers on enforcing authorities to serve safety notices in respect of products not deemed to be "*safe*". These notices may:

- prohibit the placement/supply of the product on the market (regulation 11);

- require "*dangerous*" products to be marked with warnings (regulation 12);

- require certain persons at risk to be given specific warnings, to the extent that it is practicable to do so (regulation 13);

- require the withdrawal of the product from the market, and for consumers to be alerted (regulation 14); and

- require the recall of a product (regulation 15).

These powers may only be exercised in certain conditions, with withdrawal and recall safety notices effectively functioning as 'last resort'

---

27 Section 13 CPA 1987

28 Section 13(4) CPA 1987

actions by authorities. Breach of the regulations (or a failure to comply with safety notices) is not actionable in civil law, but may result in criminal penalties[29].

In the absence of any such notice being issued, there is a lack of clear case law on whether the courts will find a failure to warn, or to recall, to be actionable in negligence. This is in direct contrast to the position in the United States and Canada, where case law on the duty to warn purchasers of defective products is clear[30].

The domestic case of *Carroll and Others v Fearon and Others*[31], concerning defective tyres, hints at the existence of a duty to warn. The Court of Appeal approved the finding of the judge below that:

> "...Dunlop failed to take "appropriate action to bring the matter to the attention if not of the owner, to the authorities concerned with road safety". They had failed to protect the public. This constituted a breach of duty."

However, negligence was not made out due to a lack of evidence on causation, specifically whether any steps taken would have led to the removal of the tyres. Accordingly, the court did not consider the parameters of such a duty of care any further.

Eric Hobbs (Farms) Limited v The Baxenden Chemical Company Limited[32] also concerned the duty to warn, albeit in an entirely different context. In 1980, a managing director of a farm had contracted for the supply and installation of insulation materials which were, on the court's findings, misrepresented as being 'self-extinguishing'. The insulation materials were, in fact, combustible, and in 1986 caused extensive fire spread and damage on the farm. The High Court placed significant importance on memoranda circulated across the manufacturer's

---

29 See regulations 42, 20, 28 and 31

30 See footnote 26

31 See footnote 22, at paragraphs 41-42.

32 [1992] 1 Lloyd's Rep. 54

business in January 1985 which, in summary, explained that such descriptions were not to be used, and that customers should be warned of the dangers associated with the insulation materials.

The court held that the contractors who were selling, or had sold, these materials "*...should have been given the same information and asked to tell past customers*". The court also remarked that announcements should have been made in the media and technical journals. The manufacturer's failure to take such steps was negligent:

> "*...a manufacturer's duty of care does not end when the goods are sold. A manufacturer who realises that omitting to warn past customers about something which might result in injury to them must take reasonable steps to attempt to warn them, however lacking in negligence he may have been at the time the goods were sold.*"

Case law on the duty to go one step further and *recall* dangerous products is equally lacking. The most authoritative case that the authors are aware of is *Wright and Cassidy v Dunlop Rubber Company Limited and Imperial Chemical Industries*[33], which concerned the supply of a chemical powder used as an antioxidant in Dunlop's tyre factory. Dunlop's employees came into contact with this power in 1946-1947, and as a result contracted bladder cancer. The defendant had been aware of the health risks facing Dunlop's employees, and had failed to warn Dunlop, despite taking steps to limit the exposure of its own employees.

Whilst the court below held that the product should have been withdrawn by 1 January 1940, the Court of Appeal did not comment specifically on this, and instead concluded that "*[the] least they should have done was to have warned Dunlops of the suspicions which they had*". The following observations were made in respect of the standard of care:

---

33 (1972) 13 K.I.R. 225

> "...the answer to the question "What are reasonable steps?" must depend upon the particular facts. It is obvious, also, that the duty is not necessarily confined to the period before the product is first produced or put on the market...
>
> the manufacturer has failed in his duty if he has failed to do whatever may have been reasonable in the circumstances in keeping up to date with knowledge of such developments and acting with whatever promptness fairly reflects the nature of the information and the seriousness of the possible consequences."[34]

The court commented that a manufacturer may be under a duty to "*cease forthwith to manufacture or supply the product*", or to take less drastic action as the circumstances justified. Thus, at an appellate level, there is no clear indication from the courts that manufacturers will be under a duty to recall dangerous products.

In the unreported case of *Walton v British Leyland UK Limited*[35], the claimant suffered injuries as a result of a wheel detaching from a car. The design of the car was defective, and the manufacturer had received reports of the defect in other cars. Whilst its press release made reference to a "*small number of rear wheel bearing failures*", the manufacturer did not recall the cars, or otherwise alert users to the extent of the dangers posed. The claimant established that the manufacturer was aware of the defect at the time of the car's sale, which in itself was sufficient to establish negligence[36]. Yet Willis J went further and expressed the view that British Leylands had a duty to:

> "...recall all cars which they could in order that the safety washers could be fitted... They seriously considered recall and made an estimate of the cost at a figure which seems to me to have been in no way out of proportion to the risks involved. It was decided not to

---

34 Ibid, pages 12-13

35 13 July 1978, High Court (Queen's Bench Division)

36 See the earlier section of this chapter entitled 'Negligence in respect of defective vehicles'

> *follow this course for commercial reasons. I think this involved a failure to observe their duty to care for the safety of many who were bound to remain at risk…"*

Notwithstanding the unequivocal nature of Willis J's comments, they are obiter dicta. The lack of subsequent case law to this effect makes it difficult to affirm that this approach would be followed by the courts in future.

Our view is that there is certainly an argument to be had that the only 'reasonable step' open to a CAV manufacturer is the recall of the product, in light of the risk of serious injury or death occurring. This will, however, depend on the nature of the defect itself: if owners or Users in Charge could install a workaround, or eliminate exposure to the defect (for example, by not activating certain features), a clear warning from a manufacturer may be enough to discharge its duty of care.

To establish causation, claimants will need to show that had a warning or recall been issued, it would have been heeded, and therefore would have prevented an accident occurring. We all know of (or may in fact be) an individual who fails to pay attention to safety warnings, safety updates, or instruction manuals; claimants in this position are likely to find it difficult to argue that the lack of warning was causative of their accident. Telematic evidence is likely to be decisive in such cases: if a manufacturer could establish that a claimant frequently ignored warnings issued by the CAV, or often failed to install safety-critical updates in a timely manner, a court may find that the claimant would not have heeded a safety warning even if it had been given (or that he or she would have failed to do so prior to the accident).

In the event that the courts do recognise such duties of care, developers and manufacturers of CAVs will need to issue safety alerts, warnings, and updates on a regular basis to discharge their duties of care. These may be easily installed in (or sent to) CAVs electronically, and without the owner or User in Charge needing to attend upon the manufacturer or a garage. But if CAVs require regular recalls (perhaps to update their hardware to accommodate new software, as we currently experience with other forms of technology), a constant cycle of recall, repair, and re-issue would render the typical model of car sales untenable. A move toward leasehold ownership would be a better fit for CAVs, enabling a swift re-issue or resolution of defects. This would also reflect current ownership trends for other items of software, which are delivered to users through a lease or licence (for example, computer software).

Negligent testing and certification

Claimants may also be turning their attention to any third parties who have tested a particular model of CAV and certified it as safe. At the time of writing there were no reported cases concerning the testing or certification of automated vehicles, but an analogy could be drawn with the case of *Perrett v Collins*[37], involving a claimant who was injured during the crash of a light aircraft. An inspector had confirmed that the aircraft was airworthy, and as a result a flying association had issued a certificate stating that the aircraft was fit to fly. The Court of Appeal recognised that the regulatory framework, including the provision of such certificates, was in place to protect the public's safety, and that the erroneous provision of a certificate would foreseeably affect aircraft passengers. Accordingly, both the inspector and the flying association were held to owe a duty of care to the injured claimant.

As a final rider, we ought to mention that where there is a contract between a claimant and a defendant (for example, a buyer and seller of a CAV), the contract may seek to exclude or limit recovery in negligence for any damages arising from a CAV's defects. Such limitations will be subject to consumer protection legislation where applicable[38], and would not affect claims brought under the CPA 1987.

---

37 [1998] 2 Lloyd's Rep. 255

38 See the Unfair Contract Terms Act 1977 and the Consumer Rights Act 2015

PRODUCT LIABILITY CASES UNDER THE AEVA REGIME

As currently drafted, the AEVA does not provide a solution to any of these product liability pitfalls: as explained in Chapter 2 ("Overview of the AEVA 2018"), the AEVA is primarily a mechanism to ensure that claimants are swiftly compensated for injury or damage caused by a CAV.

Section 6(5) of the AEVA prohibits insurers from seeking to recover their payouts through the Civil Liability (Contribution) Act 1978. Instead, Section 5(1) of the AEVA provides that *"…any other person liable to the injured party in respect of the accident is under the same liability to the insurer or vehicle owner"*. In practice, this means that any liability a manufacturer, distributor or other third party may have to a claimant in respect of a defective CAV becomes a liability owed to the insurer. An insurer would action this liability by bringing one (or more) of the claims we have explored in this chapter.

In light of all the difficulties discussed throughout this chapter, insurers may find themselves taking on risky and uncertain causes of action, and possibly failing to recover their outlay. In the absence of any change to product liability laws to resolve these problems, insurance companies will presumably have to price these risks into their policies, which may in turn prohibit a wide uptake of CAVs.

SOME FUTURE LEGAL PROBLEMS

Our view is that current product liability laws in England and Wales require reform. The smooth operation of CAVs is but one instigator: such reforms would be of broader benefit to our increasingly technology-based society.

In particular, the failure to explicitly and clearly account for software defects (save for contract claims brought under the Consumer Rights Act 2015) renders current product liability laws redundant for our modern lives. To address the problems discussed in this chapter, a reform of product liability law would, at the least, need to address:

- Application of the law to software and other electronic products or transfers, as opposed to only accommodating tangible 'goods';

- The recovery of damages for the product itself. CAVs are a good example of how significant such losses can be;

- The ability of commercial parties (e.g. the shop owner mentioned earlier) to recover for damage to private property which is caused by defective products.

Looking specifically at product liability issues concerning CAVs, there are two areas which seem particularly prone to satellite litigation which may benefit from future legislation:

- The identification of the fault within the CAV, and the subsequent identification of the correct defendant. As the law currently stands, claimants may need to protectively issue proceedings against all manufacturers or suppliers who have had some part to play in the production of the CAV, as opposed to issuing solely against the 'main' manufacturer of the CAV. Legislation could permit (and encourage) claimants to take the latter course, specifying that manufacturers will be liable for the inclusion of others' defective products, subject to having rights of recovery/indemnity;

- The standards expected of Level 3 CAVs, particularly in relation to handovers and warnings, which may otherwise result in extensive litigation.

At the time of writing, the authors are aware of an impending European Commission report into product liability laws, which will explore whether any guidance or updates to this area of law are required. This may well be the start of a dialogue on an overhaul of product liability laws, but the overall process is unlikely to be swift.

In England and Wales, an added dimension of complexity is the question of Brexit, which at the time of writing is still looming without definite decisions as to how and when the United Kingdom will leave the European Union (if it does indeed stand by its decision to do so). Assuming that Brexit occurs within the relatively nearby future, EU-instigated reform is unlikely to be taken up by the UK, unless this was required by virtue of any withdrawal agreement with the EU.

A failure to overhaul domestic product liability laws within a similar timeframe to the EU could lead to a sense of frustration on the part of insurers and consumers alike, who may feel they are unable to attain proper compensation from CAV manufacturers, and that they are unable to hold the CAV industry accountable for safety failings. This sense of frustration is likely to be exacerbated by the knowledge that the rest of the continent had enhanced protections. It is not unrealistic to suggest that this could impact the widespread uptake of CAV technology within the UK.

On the other hand, signing up to a post-Brexit EU initiative, and one which may encourage our courts to consider future case law of the CJEU, may be unpalatable. The political will may therefore necessitate a *domestic* review of product liability laws. However the changes come about, it seems inevitable that they must do so.

SUMMARY OF POINTS ON PRODUCT LIABILITY CLAIMS

The CPA 1987 is the main piece of legislation governing product liability claims
The CPA 1987 restricts recoverable losses. Of significance to CAVs is the inability to recover for commercial property damage, or for damage to the defective CAV itself
It seems unlikely that the CPA 1987 will cover defective software
The standards of safety which will be enforced by the courts are currently very uncertain, and are likely to continually shift in light of developments in CAV technology
Such issues are likely to require expert evidence, which may be difficult to obtain
Contractual claims are restricted to a narrower pool of claimants, and are likely to be limited in scope
Negligence claims may continue to play a key role in product liability law, especially if claimants issue concurrent negligence claims to protect their position
The extent to which the courts will adjudicate on negligent failures to warn users, or a failure to recall dangerous products, is yet to be seen
It is possible that those who test and certify CAVs as safe may be held accountable in negligence
The AEVA makes claimants' lives easier, but does not resolve these underlying problems
Product liability laws require reform

CHAPTER TEN
DATA AND PRIVACY
ALEX GLASSBROOK

"The algorithms of the law must keep pace with new and emerging technologies."

(Lord Justice Haddon-Cave and Mr Justice Swift, in *The Queen (on the application of Edward Bridges) v The Chief Constable of South Wales Police (Defendant), the Secretary of State for the Home Department (Interested party), the Information Commissioner and the Surveillance Camera Commissioner (Interveners)* [2019] EWHC 2341 (Admin), 4 September 2019)

INTRODUCTION

So begins the judgment of the Divisional Court as to whether the use of Automated Facial Recognition technology ("AFR") by a police force when filming crowds, in search of suspected criminals, would infringe the right to privacy of the many non-suspects whose faces are also filmed.

The lawfulness of AFR technology has just started to be determined in the courts. The use of mass data by Connected and Autonomous Vehicles, and by the linked technology of Smart Highways, is not yet part of everyday experience. Its lawfulness has yet to be adjudicated.

Data and privacy in CAV law is a complex, speculative topic. The Automated and Electric Vehicles Act 2018 does not deal with data in relation to automated vehicles (though it does provide expressly for regulations to be made in relation to energy consumption and geographical data sent by electric vehicle charging points, in Part 2 of the Act, at section 14). The Law Commission's 2018 to 2019 consultation on the law of automated vehicles[1] excluded the law of data, save as it arose in relation to the retention of accident data by automated vehicles, for use by insurance companies and litigants in AEVA 2018 claims.

Much law has yet to be written. The survey of existing laws is therefore just that: the preparation for a longer, more arduous expedition. There are, however, several emerging points of interest, even at this early stage.

OBJECTIVES OF THIS CHAPTER

This chapter:

- Provides the background to the use of data in the context of connected and automated vehicles

- Describes the laws which now apply

- Asks how the laws of data and privacy might develop in relation to connected and automated vehicles

BACKGROUND TO DATA, PRIVACY AND CONNECTED AND AUTOMATED VEHICLES

Connected and automated vehicles will be fuelled, in a sense, by data. Information as to the location of the vehicle, its desired destination and even the driver's health (from sensors in the steering wheel), is processed by the connected vehicle and its systems.

Even now, surveillance systems on motorways allow highway authorities to plot the route of a particular vehicle, by recognising its number plate, and to calculate from its progress (and the progress of other vehicles) the speed limits that it might introduce further along the route, in order to maintain smooth traffic flow.

---

1 See https://www.lawcom.gov.uk/project/automated-vehicles/

Highway management systems will become more sophisticated and widespread, with the introduction of Smart Highways, 5G networks and the greater prevalence of Connected and Autonomous Vehicles (CAV's) which are able to communicate both with each other and with road infrastructure ("V2V" and "V2I" connectivity).

Such connectivity has been said, by the UK Department of Transport, to be a key part of its strategy for road transport[2].

The Place of Insurance in the Law of Data relating to Connected and Autonomous Vehicles

The law of liability relating to motor vehicles rests, in the United Kingdom, upon compulsory insurance for the victims of accidents[3]. Part 1 of the Automated and Electric Vehicles Act 2018 (discussed in the second part of this book) continues that tradition and is a part of it.

In the context of a liability system rooted in compulsory insurance, the gathering of accident information (to satisfy the insurer of the fact of an accident falling within that insurance scheme), is key.

Part 1 of the AEVA 2018 increases insurers' potential liabilities in the case of AV accidents, via the strong remedy in section 2 (discussed in previous chapters). At the same time, the technology of AVs provides insurers with much more information in relation to any accident, via the systems of the car, including cameras.

---

[2] See page 60 of the Department of Transport Investment Strategy (December 2017) at https://assets.publishing.service.gov.uk/government/uploads/system/uploads/attachment_data/file/624990/transport-investment-strategy-web.pdf

[3] For an account of the evolution of the law of road traffic accident liability in the UK, and how compulsory third party insurance came to be at its core, see Chapter 1 of The Law of Motor Insurance by Professor Robert Merkin QC and Maggie Hemsworth (Sweet and Maxwell, 2nd edition, 2015) and paragraphs 1 to 5 of the judgment of Lord Sumption in *Cameron v Liverpool Victoria Insurance Company Limited* [2019] UKSC 6.

So CAVs will alter features of the legal landscape: not only in relation to liability, but also in relation to data, privacy and insurance law.

Data, CAVs and Criminal Law

As Chapter 11 discusses, connected and autonomous vehicles will also change road transport law in the criminal field. CAV data will expand and change the nature of criminal investigation, charge and trial.

The *Bridges* case (above) is likely to proceed further, on appeal, and will tell us a great deal about the use of mass data and surveillance and its effects upon individuals' data rights.

The Broader Commercial Aspects of Data and Privacy in Future Transport

There are, on the other hand, arguments in favour of individual privacy. Insurance is not the only business which is likely to profit from vehicles in their future form. New, connected and automated vehicles are likely to open new avenues where businesses might thrive. And those businesses are likely to nourish themselves largely upon data.

An automated, driverless pod might be summoned from an App, for example. App car businesses will be backed by insurance, but liability for accidents might not be their only motive to gather data. Data gathered during your journey (as to destinations, frequency of journeys, online shopping preferences etc) will have a value, and might be sold on.

Those are matters outside the immediate scope of this book, which is mainly concerned with liabilities arising from the familiar and unwelcome major legal effect of road transport today: injury through road traffic accidents.

But the unwelcome effects of road transport in the future might be different. We hope that a beneficial effect of connected and automated vehicles will be to reduce road traffic accident casualties. But, if so, might we notice other disadvantages of easy transport? Might we pay greater attention to the use – and value – of our data?

THE LAWS OF DATA AND PRIVACY AS THEY RELATE TO CONNECTED AND AUTOMATED VEHICLES

Current Laws

The **Human Rights Act 1998** gives effect in British law to certain rights in the European Convention on Human Rights (ECHR). The rights are listed in Schedule 1 of HRA 1998, and include:

- The right to life (Article 2 ECHR)

- The right to liberty and security (Article 5)

- The right to a fair trial (Article 6)

- The right to respect for "*private and family life, … home and correspondence*" (Article 8). "*There shall be no interference by a public authority with the exercise of this right except such as is in accordance with the law and is necessary in a democratic society in the interests of national security, public safety or the economic well-being of the country, for the prevention of disorder or crime, for the protection of health or morals, or for the protection of the rights and freedoms of others*" (Article 8(2)).

- The right to freedom of expression, including "*freedom to hold opinions and to receive and impart information and ideas without interference by public authority and regardless of frontiers*" (Article 10).

- "*The enjoyment of the rights and freedoms set forth in this Convention shall be secured without discrimination on any ground such as sex, race, colour, language, religion, political or other opinion, national or social origin, association with a national minority, property, birth or other status*" (Article 14, the prohibition of discrimination).

- Some ECHR rights are absolute (eg. the prohibition on torture under Article 3, which one hopes will not arise in the CAV context) while some (eg. the right to respect for private and family life under Article 8) are qualified, so might be interfered with by the State (see Article 8(2) above). However, "*The restrictions permitted under this Convention to the said rights and freedoms shall not be applied for any purpose other than those for which they have been prescribed*" (Article 18, Limitation on use of restrictions on rights).

(That is not a complete list of the ECHR rights, but a selection of those rights which seem potentially applicable to CAVs in relation to data and privacy laws.)

"*It is unlawful for a public authority to act in a way which is incompatible with a Convention right*" (section 6(1) HRA 1998). A "*public authority*" includes "*(a) a court or tribunal, and (b) any person certain of whose functions are functions of a public nature …*", although "*In relation to a particular act, a person is not a public authority by virtue only of subsection (3)(b) if the nature of the act is private*" (section 6(5)).

"*So far as it is possible to do so, primary legislation and subordinate legislation must be read and given effect in a way which is compatible with the Convention rights*"
(section 3(1) HRA 1998, "Interpretation of legislation").

An award of damages may be made for breach of those rights if, having regard to other remedies, the court "is satisfied that the award is necessary to afford just satisfaction to the person in whose favour it is made" (section 8(3), "Judicial remedies").

The **Data Protection Act 2018 (DPA 2018)** enacts the European Union's General Data Protection Regulation 2016 (the GDPR). The British government committed to implementing the GDPR, irrespective of the result of the 2016 British referendum on membership of the EU, and the GDPR was given force in British law by the DPA 2018. That decision was, no doubt, influenced by the extra-territorial effect of the GDPR (applying to anyone doing business in the EU, regardless of EU membership).

The relevant parts of the DPA 2018 are as follows. This is not a comprehensive account of the workings of the Act, but a focus upon those areas of the DPA 2018 which seem most likely to affect CAVs.

As section 1 DPA 2018 ("Overview") states:

> "(1) This Act makes provision about the processing of personal data.
>
> (2) Most processing of personal data is subject to the GDPR.
>
> (3) Part 2 supplements the GDPR (see Chapter 2) and applies a broadly equivalent regime to certain types of processing to which the GDPR does not apply (see Chapter 3)."

Both the GDPR and DPA 2018 are of broad application, and the latter (by invoking the former) sets strict rules for how personal information is dealt with. The following parts of section 3 ("Terms relating to the processing of personal data") show the breadth of the Act:

> "*3 Terms relating to the processing of personal data*
>
> (1) This section defines some terms used in this Act.
>
> (2) "Personal data" means any information relating to an identified or identifiable living individual …

(3) *"Identifiable living individual"* means a living individual who can be identified, directly or indirectly, in particular by reference to—

(a) *an identifier such as a* name, *an* identification number, *location data or an* online identifier, *or*

(b) *one or more factors specific to the physical, physiological, genetic, mental, economic, cultural or social identity of the individual.*

(4) *"*Processing*", in relation to information, means an* operation or set of operations *which is* performed on information, *or on sets of information, such as—*

(a) *collection, recording, organisation, structuring or storage,*

(b) *adaptation or alteration,*

(c) *retrieval, consultation or use,*

(d) *disclosure by transmission, dissemination or otherwise making available,*

(e) *alignment or combination, or*

(f) *restriction, erasure or destruction,*

...

(5) *"*Data subject*" means the* identified *or* identifiable living *individual to* whom personal data relates.

...

(7) *"Filing system" means any structured set of personal data which is accessible according to specific criteria, whether held by automated*

means or manually and whether centralised, decentralised or dispersed on a functional or geographical basis.

…"

The DPA 2018 defines some terms in a complicated way, not by a single definition but by reference to the parts of the Act in which they are treated. It defers to a large extent to the terms of the GDPR itself, where the definition of terms is often clearer. For example, data controller and data processor are defined in Article 4 GDPR as follows:

> "*(7) 'controller' means the natural or legal person, public authority, agency or other body which, alone or jointly with others, determines the purposes and means of the processing of personal data;* …
>
> *(8) 'processor' means a natural or legal person, public authority, agency or other body which processes personal data on behalf of the controller;*"

The GDPR 2016 and DPA 2018 require that personal data be processed in accordance with the following **six data protection principles**. GDPR 2016, Article 5(1), and sections 34 to 40 DPA 2018, in Chapter 2, set out the six data protection principles. section 34(3) DPA 2018 provides that "*The controller in relation to personal data is responsible for, and must be able to demonstrate, compliance with this Chapter*".

The six data protection principles are:

1. **Lawfulness, fairness and transparency** (in relation to the data subject)

2. **Purpose Limitation** (limitation to the purposes of processing the data, which purposes must be specified, explicit and legitimate)

3. **Data Minimisation** (of processed data, to achieve the purpose)

4. **Accuracy** (of data, including the erasure of outdated information)

5. **Storage Limitation** (limitation of storage period to the time necessary for the purpose)

6. **Integrity and Confidentiality** (ie. secure storage)

The lawful purposes for the processing of personal data are set out in Article 6(1) GDPR 2016:

"(a) the data subject has given consent to the processing of his or her personal data for one or more specific purposes;

(b) processing is necessary for the performance of a contract to which the data subject is party or in order to take steps at the request of the data subject prior to entering into a contract;

(c) processing is necessary for compliance with a legal obligation to which the controller is subject;

(d) processing is necessary in order to protect the vital interests of the data subject or of another natural person;

(e) processing is necessary for the performance of a task carried out in the public interest or in the exercise of official authority vested in the controller;

(f) processing is necessary for the purposes of the legitimate interests pursued by the controller or by a third party, except where such interests are overridden by the interests or fundamental rights and freedoms of the data subject which require protection of personal data, in particular where the data subject is a child.

Point (f) of the first subparagraph shall not apply to processing carried out by public authorities in the performance of their tasks."

DATA AND PRIVACY • 163

Furthermore, special rules apply to sensitive ("special") personal data. Article 9 of the GDPR 2016 defines *"special categories of personal data"*, processing of which is prohibited unless a prescribed situation applies.

The special categories of personal data of which processing is prohibited under Article 9, unless specifically permitted, are:

> *"personal data revealing racial or ethnic origin, political opinions, religious or philosophical beliefs, or trade union membership, and the processing of genetic data, biometric data for the purpose of uniquely identifying a natural person, data concerning health or data concerning a natural person's sex life or sexual orientation"*
>
> (Article 9(1))

[handwritten: Location data could reveal such things]

The exceptional circumstances in which processing of those sensitive personal data is permitted by Article 9 are as follows:

> *"(a) the data subject has given explicit consent to the processing of those personal data for one or more specified purposes, except where Union or Member State law provide that the prohibition referred to in paragraph 1 may not be lifted by the data subject;*
>
> *(b) processing is necessary for the purposes of carrying out the obligations and exercising specific rights of the controller or of the data subject in the field of employment and social security and social protection law in so far as it is authorised by Union or Member State law or a collective agreement pursuant to Member State law providing for appropriate safeguards for the fundamental rights and the interests of the data subject;*
>
> *(c) processing is necessary to protect the vital interests of the data subject or of another natural person where the data subject is physically or legally incapable of giving consent;*
>
> *(d) processing is carried out in the course of its legitimate activities with appropriate safeguards by a foundation, association or any other not-for-profit body with a political, philosophical, religious or trade*

union aim and *on condition that the processing relates solely to the members or to former members of the body* or to persons who have regular contact with it in connection with its purposes and that the personal data are not disclosed outside that body without the consent of the data subjects;

(e) processing relates to personal data which are manifestly made public by the data subject;

(f) processing is necessary for the establishment, exercise or defence of legal claims or whenever courts are acting in their judicial capacity;

(g) processing is necessary for reasons of substantial public interest, on the basis of Union or Member State law which shall be proportionate to the aim pursued, respect the essence of the right to data protection and provide for suitable and specific measures to safeguard the fundamental rights and the interests of the data subject;

(h) processing is necessary for the purposes of preventive or occupational medicine, for the assessment of the working capacity of the employee, medical diagnosis, the provision of health or social care or treatment or the management of health or social care systems and services on the basis of Union or Member State law or pursuant to contract with a health professional and subject to the conditions and safeguards referred to in paragraph 3;

(i) processing is necessary for reasons of public interest in the area of public health, such as protecting against serious cross-border threats to health or ensuring high standards of quality and safety of health care and of medicinal products or medical devices, on the basis of Union or Member State law which provides for suitable and specific measures to safeguard the rights and freedoms of the data subject, in particular professional secrecy;

(j) processing is necessary for archiving purposes in the public interest, scientific or historical research purposes or statistical purposes in accordance with Article 89(1) based on Union or Member State law which shall be proportionate to the aim pursued, respect the essence of the right to data protection and provide for suitable and specific measures to safeguard the fundamental rights and the interests of the data subject."

And Article 9(3) adds that:

"Personal data referred to in paragraph 1 may be processed for the purposes referred to in point (h) of paragraph 2 when those data are processed by or under the responsibility of a professional subject to the obligation of professional secrecy under Union or Member State law or rules established by national competent bodies or by another person also subject to an obligation of secrecy under Union or Member State law or rules established by national competent bodies."

The GDPR 2016 further provides that:

- The GDPR does not apply in all circumstances. In particular, it does not apply to the processing of personal data "*(c) by a natural person in the course of a purely personal or household activity*" or "*(d) by competent authorities for the purposes of the prevention, investigation, detection or prosecution of criminal offences or the execution of criminal penalties, including the safeguarding against and the prevention of threats to public security*" (Article 2).

- Data controllers and processors must show how they comply with all of the data protection principles. This is the overarching "*accountability*" requirement (Article 5(2).

- Data protection must be achieved by technical design and by default (Article 25).

- The territorial reach of the regulation is wider than the previous law. It applies to data controllers and processors "*offering goods and services to data subjects who are within the Union*", even

- where the controller or processor is established outside the EU (see the preamble at paragraph 23 and Article 3).

- The meaning of "*personal data*" is wider than previously understood. For example, information identifying online activity, such as an IP address, can be personal data, as can data that has been "*pseudonyomised*" (see preamble at 26, 30 and Article 32.

- Consent to data processing must be "*given by a clear affirmative act*" and "*unambiguous*". Silence or not altering a pre-ticked box will not suffice. Consent must be capable of withdrawal at any time (preamble at 32 and Article 7).

- Information to data subjects should be "*easily accessible and easy to understand*" (preamble at 58 and Article 12).

- The data subject has a right to have data provided in intelligible form to him or her and to transmit it to another controller, or where technically feasible to ask a data controller to transmit that data ("*data portability*" – preamble at 68 and Article 20)

- Controllers and processors owe enhanced duties to ensure data security, by several measures including (for example) "*pseudonymising*" personal data as soon as possible (preamble at 78)

- Where there is a "*high risk*" in data processing "*to the rights and freedoms of natural persons*", the controller shall carry out "*an assessment of the impact of the envisaged processing operations on the protection of personal data*". That is especially so "*where a type of processing*" uses "*new technologies*" (Article 35)

- Member states may "*provide for more specific rules to ensure the protection of the rights and freedoms of in respect of the processing of employees' personal data*" (Article 88).

- The GDPR states its commitment to a balance between the data subject's security and the demands of increasingly sophisticated technologies and international data transfers. But it gives greater weight to the data security of the individual subject (see preamble at 6 and 7). That is its object.

In the UK, the Court of Appeal – *W M Morrison Supermarkets Plc v Various Claimants* [2018] EWCA Civ 2339 – held that an employer was (on the facts of that case) vicariously liable for the criminal act of an employee who had disclosed personal and confidential information of other employees online, in breach of the preceding British data protection statute, the Data Protection Act 1998. At the time of writing, a further appeal is due to be heard by the UK Supreme Court.

Finally, the **Protection of Freedoms Act 2012** (PFA 2012) provides for a code of practice in relation to camera surveillance. PFA 2012 arose, in the context of mass data gathering for law enforcement purposes, in the *Bridges* case, discussed in this chapter. It is relevant in the vehicles context because it applies to Automatic Number Plate Recognition (ANPR) technology (section 29(6)(a)). ANPR is used as a tool not only of law enforcement but also of traffic management. So PFA 2012 has a potential relevance to connected and autonomous vehicles.

The dynamics of the PFA 2012 argument, were it to arise in a CAV case, would depend upon the facts before the court so are unpredictable. But the appearance before the courts of ANPR, a component of present and (probably to a greater extent) future traffic-management technology, brings the PFA 2012 at least into our peripheral legal vision.

CAVs and Mass Data Processing: An Analogous, Recent Judgment on Mass Data Use & Privacy: Automated Facial Recognition

As the judgment in *Bridges*, the Automated Facial Recognition (AFR) case, notes, the parties brought the proceedings before the court "*to seek the court's early guidance as regards the legal parameters and framework relating to AFR, whilst it is still in its trial phase, and before it is rolled-out nationally*"[4].

The case arose in the context of law enforcement, and specifically the facial identification by concealed, technological means, of criminal suspects within a crowd which included many non-suspects. So the context differs from the use of mass data in road transport. But the case provides a good, recent example of how a British court dealt with the legality of a mass data operation by the State (noting, of course, that it was not a final decision in the sense that it might yet be appealed).

AFR was noted not only to involve the taking of photographs but also detailed analysis by software of biometric data, of any subject photographed (which was key to the Divisional Court's findings that data had been processed and the Claimant's right to privacy under Article 8 of the European Convention on Human Rights had at least been interfered with, if not unlawfully breached[5]).

The judgment of the Divisional Court was that, while the non-suspect Claimant's data protection rights and rights to privacy were engaged in law, existing laws and regulations governing the police's use of AFR provide sufficient protection of privacy, and that AFR was used lawfully in the trial period. The court noted, in particular, that in its view:

> "*Any interference with the Claimant's Article 8 rights would have been very limited. The interference would be limited to the near instantaneous algorithmic processing and discarding of the Claimant's biometric data*"[6].

The court concluded that "the current legal regime is adequate to ensure the appropriate and arbitrary use of AFR Locate and that [South Wales Police's] use to date of AFR Locate has been consistent with the

---

4 Paragraph 2 of the judgment in *R (ex p Bridges) v Chief Constable of South Wales Police* [2019] EWHC 2341 (Admin).

5 See paragraph 54 of *Bridges*.

6 Paragraph 101 of *Bridges*.

requirements of the Human Rights Act, and the data protection legislation"[7].

That conclusion might yet be challenged on appeal.

Bridges is an interesting case in our context, due to the legal issues (data and privacy) and the factual context (an innovative, artificially intelligent tool, used in relation to a mass of subjects).

The *Bridges* case also highlights a gap in the current legal preparation for CAVS:

The Gap: No current British Regulations nor Code of Practice Specifically in relation to the Protection of Data arising from the use of Connected and Autonomous Vehicles

The prospect of mass processing of data by artificially intelligent machines raises the anxiety of breaches of privacy. There are reminders of this even in the language of the ECHR rights. Article 8, for example, expressly protects *"everyone's … correspondence"*, and an inevitable feature of connected and autonomous vehicles will be their ability to allow occupants to email, text, chat, etc using the systems of the vehicle. The GDPR 2016 is, in a sense, an updated and more detailed version of an earlier, express right.

The process of safeguarding mass data raises challenging questions of human and machine behaviours, and of the ethics of the use of data by the state and by business.

Mass data gathering will be a feature of connected and automated vehicles. Individual vehicle users (whether as drivers or passengers) will provide their data both to the state and to commercial concerns.

Highway authorities are likely to find data a useful tool to prevent road congestion and improve road safety. Vehicle insurers (especially where

---

7 Paragraph 159 of *Bridges*.

subject to the strong liability of the Automated and Electric Vehicles Act 2018) will wish to be furnished with the best available data about prospective risks and, in the event of an accident, of all the information relating to all vehicles' speed, position etc leading to and at the moment of collision, at the earliest opportunity.

Vehicle technology already incorporates many sensors, including cameras, and future vehicle technology is likely to provide more sources of information.

Despite those concerns, the writer is aware of no Code of Practice in the UK in relation to data protection issues arising from CAVs. The AEVA 2018 does not legislate for data rights in relation to automated vehicles.

The closest, current equivalent to a Code, of which the writers are aware, is the April 2018 paper of the "Berlin Group" (the International Working Group on Data Protection in Telecommunications), which appears on the website of the International Conference of Data Protection and Privacy Commissioners (the ICDPPC)[8]. The Information Commissioner's Office (ICO) of the UK is an accredited member of the ICDPPC[9].

The Berlin Group paper identifies a number of data and privacy risks of CAVs. It categorises those risks as follows:

- **Lack of transparency**: users of CAVs might not be made sufficiently aware of the privacy risks, eg. that the camera of a vehicle might record images of people walking around the vehicle.

---

[8] https://icdppc.org/the-icdppc-working-group-on-data-protection-in-telecommunications-adopts-a-working-paper-on-connected-vehicles/ Also on the Berlin Group's site, at https://www.datenschutz-berlin.de/fileadmin/user_upload/pdf/publikationen/working-paper/2018/2018-IWGDPT-Working_Paper_Connected_Vehicles.pdf

[9] https://icdppc.org/participation-in-the-conference/list-of-accredited-members/

- **Unlawful processing**: there is a risk that data controllers might not understand the law governing data, and could therefore process data unlawfully.

- **Unauthorised secondary use**: data collected from CAVs by others could be used for purposes not authorised by the user of the vehicle (eg. commercial or criminal use).

- **Excessive collection**: CAV technologies (eg. sensors, machine learning algorithms) might lead to the unintended collection of excessive amounts of data.

- **Lack of Control**: Insufficient information or options for the user of a CAV might leave their data insufficiently protected (eg. without an easy means of erasing their data from a hired or shared vehicle, their data might remain accessible in that vehicle).

- **Inadequate security**: The quantity of different features on a CAV (entertainment, links to occupants' smartphones etc) leaves too many entry points (or "attack surfaces") through which a hacker (a data thief) can gain access.

- **Lack of accountability**: another disadvantage of the plurality of features on a connected vehicle. The plurality of service providers risks a proliferation of roles and of different understandings of data law. In case of a breach, the user might have difficulty identifying the fault and the responsible party.

The Berlin Group's suggested solutions centre upon "privacy by design", namely building data and privacy protection measures into the design of the vehicles systems – both its hardware (eg. sensors) and software (eg. allowing blurring of faces and of registration plates).

The need for such measures might be determined, in the CAV context, by the courts in a future case similar to the recent AFR case of *Bridges*.

The context of that case, however, was law enforcement. Context, of course, is everything.

HOW MIGHT DATA AND PRIVACY LAW DEVELOP THROUGH THE USE OF CONNECTED AND AUTONOMOUS VEHICLES?

That is a question for the further future, in a world of at least Level 4 vehicles, in which Part 1 of the AEVA 2018 has been in operation for some years and its principles (of causation, contributory negligence etc, discussed in part 2 of this book) have been tested and developed by the courts.

But we can mark out some likely issues:

Mass Vehicle Data and Privacy

As the judgment in the *Bridges* case shows, (and the Annex to the judgment, in which the relevant laws and Codes of Practice are extracted), technology fits within a legal framework, which includes regulations and codes of practice.

The authors are unaware of any current Code of Practice relating specifically to the use of data in connected vehicles and smart highways in the UK. We recognise that those are developing technologies. To the best of the writers' knowledge, the Berlin Group Paper of April 2018 (which is more akin to a risk assessment) is the closest equivalent.

But those technologies will, possibly in the near future, be deployed in a local trial, just as AFR was deployed before proceedings concerning its legality were brought before the court in the *Bridges* case. The legal boundaries of data use in an era of connected vehicles and smart highways might need to be tested in a similar case.

Human Supervision

A key factor in the legal use of AFR (as the Divisional Court found it to be, in *Bridges*) was that, in the event of a match being identified by the software, between the image of a person filmed and that of a criminal suspect, "the two images are reviewed by an AFR operator ("the system operator", who is a police officer) to establish whether he believes that a match has in fact been made". In the court's view, "the fact that human eye is used to ensure that intervention is justified is an important safeguard"[10].

That point touches upon an issue in CAV law, which as things stand is purely speculative: is it possible that CAV technology will never progress to Level 5 (no human supervision needed) but only so far as Level 4? If so, will CAV technologies only ever be a "guardian angel" type of technology (in the writer's phrase), assisting the human driver? Or is that an overly anthropomorphic view? That is entirely a matter for imagination, rather than legal comment, so I leave that point for a future edition.

Car Sharing

There has been discussion online of the issue of car-users' data, contained on smartphones, which is shared with the systems of a rental or shared car. The question arises as to whether this is covered by the DPA 2018 and, if so, what are the requirements upon the controller of the data.

The answer will depend upon the facts. In the first place, the question would arise as to whether or not data had been processed, which would involve understanding the working of the systems of the car in question. In the case of a purely domestic car share, it might be argued that the data protection laws have no application (see GDPR Article 2(c), above), although that would require a "purely personal or household acitivity". The writer offers no general answer, but notes that the Berlin

---

10 Paragraph 33 of the Divisional Court's judgment in *Bridges* (above).

Group paper (see above) raised the issues including transparency of systems and of erasure of data.

CAVs and Sensitive Data

"An app for everything" is an exciting phenomenon, but it generates its own problems. That is particularly true of data problems. Software depends upon data to operate. Apps tend to offer convenience to the user, by carrying out mundane or everyday tasks. But, to do so, an App usually needs access to the user's personal information. And that might include sensitive information.

In the vehicle context, a good example is the use of health information. Even at proximate levels of vehicle automation, "guardian angel" devices (in the writer's phrase) are being publicised. Several devices aim to detect if the driver has lost concentration, for example by falling asleep (or even, according to a recent report, by becoming angry). Those devices monitor physiological signs (eg. breathing, heart rate), through sensors in the car (eg. in the steering wheel). That generates sensitive information, as to health. And that information must, if it is to be processed at all, be given special data protections (see above).

The same might be true of destination information: a particular destination might imply a religious or political belief.

CONCLUSION

The current debate on data in CAV law is as to the retention of accident data, within existing data protection laws (above). Insurance companies press, understandably, for a full disclosure of accident data stored by a vehicle in the period before and after a collision, as well as at the moment of the accident. The provision of that data would, in principle, be helpful to all parties in resulting litigation, and to the court.

But a wider study of data protection law shows how carefully the boundaries of that disclosure – and of any handling of data – must be set. Bare information often implies more sensitive facts. Destination information might imply a health problem, or a religious or political belief. The sensitivity of data boundaries is a factor of which all those with responsibility for data – especially state and commercial concerns – should be aware.

It is likely that the GDPR will not be the final word on data security relating to vehicles. If connected and autonomous vehicles evolve as rapidly as many expect, the scale on which personal data is processed will increase significantly.

With that, the demands for even more specific data security laws and practices will expand. Those laws must be practically enforceable. That will affect not just the law but also the commercial dynamics of the CAV industry, including insurance.

Court procedures will also require amendment, as the quantity of digital documents increases. The courts – and lawyers – will need to move away from a paper practice and towards e-working, to be able to navigate (and securely handle) data contained in case documents.

Our current data protection laws are – quite properly – a reaffirmation of privacy. And a forceful one at that. This is an area of very strict liability.

SUMMARY OF POINTS ON DATA AND PRIVACY

Law established since 2000 (with the coming into force of HRA 1998) gives everyone a right to respect for privacy and family life, which includes their correspondence.
Data protection laws set demanding requirements for the handling of personal data, including prohibitions and high penalties.
Data protection laws have already generated a large quantity of legislative material as well as Codes of Practice. Data protection law in the CAV field is very likely to follow that pattern.
There is uncertainty as to precisely how data protection law might develop in this field.
This is an area of strict liability and one of which businesses and government should be especially aware

CHAPTER ELEVEN
CRIMINALITY

EMMA NORTHEY

INTRODUCTION

The AEVA 2018 does not deal with any of the potential criminal offences connected with the use of CAVs on our roads. To a certain extent the existing criminal law will be applicable because CAVs are motor vehicles, but there will also be new types of criminality (as with any new technology), which may require new powers and offences to tackle them.

In its Background Papers to the Preliminary Consultation Paper[1], the Law Commission identified over 70 separate criminal offences that regulate the use of standard motor vehicles. A full examination of all the issues raised by each of those offences is beyond the scope of this work, but some of the broad themes and principles merit consideration.

OBJECTIVES OF THIS CHAPTER

The objectives of this chapter are to:

- consider which areas of the current criminal law will translate to CAVs;
- explore the types of offence that may cease to be relevant as CAVs are introduced;
- assess whether any new offences may need to be created in connection with CAV usage.

---

1 https://s3-eu-west-2.amazonaws.com/lawcom-prod-storage-11jsxou24uy7q/uploads/2018/11/Compiled-background-paper-document.pdf

THE APPLICATION OF EXISTING OFFENCES TO CAVs

As noted above, there is already an extensive body of criminal law relating to the use of standard motor vehicles. Many of these offences will be equally applicable to CAVs and will translate to their use with no more than minor adjustments, such as the clarifying of definitions. Four examples of everyday motoring offences that are likely to continue to have some application to Level 3 vehicles and, potentially, to Level 4 and 5 CAVs are discussed below.

<u>Driving while disqualified</u>

This is an offence that will continue to be relevant for as long as motor vehicles have a human driver i.e. until Level 5 CAVs come into service. As noted in Chapter 7, it is unclear at present whether operating a Level 3 or Level 4 vehicle will require additional driver training, testing and licensing. It is possible that such vehicles will become a separate category, which can be added to an existing standard licence upon the holder meeting certain criteria. It is also possible that the new role of User in Charge of a vehicle in automated mode may, if adopted, require a separate licence.

Although some amendments to the current law may be needed to take account of such changes, there is no obvious reason to interfere with the underlying principle that there should be a system for the licensing of drivers (and/or Users in Charge). Rather, it should be possible to adapt the existing system, including the criminal offences and sanctions that encourage compliance with it, to encompass any new types of licence.

It will also be necessary (in respect of this offence and the many others that are only committed when the vehicle is being driven) to keep the legal definition of the term "driving" under review, both to ensure that there is no loophole through which, say, a person, who should have been in control of a Level 3 vehicle, can escape by asserting that he was not driving because the vehicle was doing all the work, and to prevent a person, who had no control over a Level 4 CAV in automated mode, from being pursued as its driver for acts or omissions that were not his responsibility.

Driving without insurance

This is a serious offence when committed in relation to a standard vehicle, partly because it makes the life of any person, who suffers injury or property damage as a result of the uninsured driver's negligence, very much more difficult and uncertain. However, at least the MIB will usually step in to ensure any claim is paid.

Failing to insure a CAV that is listed by the Secretary of State under Section 1 of the AEVA 2018 is arguably even more reprehensible, particularly from the point of view of the victim of any road accident. As noted in Chapter 8, the "strong" remedy provided by the AEVA 2018 will not be available in such cases and, if the CAV was being used in automated mode when the accident occurred, there is also no MIB "safety net".

As with driver licensing, there does not appear to be any reason to alter the principle of compulsory insurance for all motor vehicles, including CAVs. However, if it remains the case that the victims of uninsured CAVs find it more difficult to obtain compensation, there would be strong arguments for reflecting that in the Sentencing Guidelines by specifying the fact that the uninsured vehicle was a CAV as an aggravating factor.

Driving while under the influence of drink/drugs

This is another offence that is only likely to remain relevant while there is still a human, who can be said to be "driving" the vehicle i.e. prior to the introduction of Level 5 CAVs.

One of the main advantages of a fully driverless vehicle is that it will be able to transport people who could not otherwise transport themselves (including the very young, the very old, people with certain disabilities, and people who are temporarily incapacitated through tiredness, illness or intoxication). Going home in a Level 5 CAV after overdoing it at a party could become the very model of responsible behaviour.

The temptation for an intoxicated person with access to a Level 3 or Level 4 vehicle to attempt to rely on its systems to get him home is obvious. However, the fact that such systems will only hand back control when they cannot cope with the driving task arguably increases the need for the driver to be fully capable. As above, this is something that could perhaps be reflected in the Sentencing Guidelines, with driving a Level 3 vehicle while intoxicated being an aggravating factor and therefore potentially attracting an even higher penalty than driving a standard vehicle in that condition.

Speeding

In a fully automated world, there would be no need for a single fixed speed limit. Each CAV would "know" its own limitations and those of its human passengers. It would also understand its environment: the weather conditions, the volume of traffic, the road surface etc. Using all that information, it would be able to calculate and travel at the maximum safe speed (the word "safe" could be given an extended meaning here to ensure that the speed was also fuel efficient and thus environmentally friendly).

However, while CAVs and standard vehicles continue to co-exist, the current "bright line" speed limits are likely to remain in force. It would clearly be possible to program any Level 4 or 5 CAV not to exceed the speed limit while operating in automated mode. It may also be necessary to limit the speeds at which Level 3 vehicles can travel, in order to ensure that the human driver has sufficient reaction time during a hand back of control. In the event that the speed limit becomes part of certain vehicles' programming, it would make sense to replace the offence of speeding with an offence of disabling the speed limiting software. Then, instead of having to wait for a vehicle to be caught speeding in order to take action, the authorities would have the opportunity to identify those users whose intention was to speed and to stop them before any unsafe use had occurred.

In the 2018 Preliminary Consultation, the Law Commission asked whether CAVs should ever be permitted to exceed the speed limit. The responses were mixed with strong views expressed on both sides of the argument. A temporary exception for safety reasons e.g. to accelerate out of the way of an oncoming vehicle that has lost control, seems the measure most likely to gain acceptance.

NEW FORMS OF CRIMINAL CONDUCT IN CONNECTION WITH CAV USAGE

The "connected" and "autonomous" nature of CAVs mean that they have two new features, over and above those found in standard vehicles, which those with malign intent may try to exploit. The criminal law is regularly required to adapt to new ways of committing a crime (something that it typically does relatively well) and to new forms of misconduct (which may be harder to address effectively with the existing rules and sanctions). The potential for both physical and virtual interference with CAVs and their environment, and the ability of the criminal law to respond, is considered below.

Physical interference with CAVs and their environment

CAVs depend on their sensors to tell them where they are in relation to the surrounding environment and what that environment consists of: the road, the pavements, buildings, other vehicles, pedestrians and animals. In addition, a CAV may receive information from certain elements within that environment, such as: smart pedestrian crossings, smart CCTV cameras monitoring the volume of traffic, and other CAVs.

Altering the sensors, signage or other roadside equipment upon which a CAV relies would be highly likely to alter its behaviour, bringing the vehicle to a halt or causing an accident.

It is already an offence under Section 22A of the Road Traffic Act 1988 (RTA 1988) to interfere with a motor vehicle or (directly or indirectly) with traffic equipment, in such circumstances that it would be obvious to a reasonable person that to do so would be dangerous. The term "*traffic equipment*" is widely defined and is likely to encompass signs and equipment. The position in respect of road markings is less clear because, although the same section also prohibits causing anything to be on or over a road, it is not clear that it would extend to the removal of road markings (which CAVs may be relying on to maintain lane discipline or to stop in the correct place at a junction). Such conduct might be dealt with as criminal damage, but a minor amendment to this section might be preferable to ensure that the relevant legal controls were all in one place. It would also seem sensible to extend the qualifier "*(directly or indirectly)*" to interference with a motor vehicle.

However, even with those (or similar) changes, the offence is somewhat limited in its utility by the maximum penalty, which, on conviction on indictment is a term of 7 years' imprisonment and/or an unlimited fine. This does not appear adequate for dealing with conduct that results in serious injury or death. One option would be simply to increase the maximum penalty; an alternative approach, which was put forward in the Law Commission's 2018 Preliminary Consultation[2], and which received broad support, would be to introduce a new aggravated offence of causing death or serious injury by wrongful interference with roads, vehicles or traffic equipment.

Another way in which CAVs may be physically interfered with is by harassment from other drivers and pedestrians. There have already been several reports of such behaviour during the testing of CAV prototypes of roads.

It is already an offence under Section 3 of the RTA 1988 to drive a vehicle without reasonable consideration for other persons using the road, and this may well be sufficient to address harassment of CAVs by

---

2 The Law Commission's Consultation Paper No. 240 "Automated Vehicles: A Joint Preliminary Discussion Paper", dated 8 November 2018, at paragraphs 7.101-7.103: https://s3-eu-west-2.amazonaws.com/lawcom-prod-storage-11jsxou24uy7q/uploads/2018/11/6.5066_LC_AV-Consultation-Paper-5-November_061118_WEB-1.pdf

other drivers. In addition, it is an offence under Section 137 of the Highways Act 1980 to wilfully obstruct free passage along a highway. While this would be likely to cover harassment of CAVs by pedestrians simply standing in front of them, the Law Commission note that it is less clear that it would apply to a pedestrian, who steps out in front of a CAV in order to force it to make an emergency stop[3]. In the event that this becomes a popular pastime, which is entirely possible given young people's fascination with risk-taking (playing "chicken" with cars and trains, taking "selfies" from the top of cranes etc), a new offence may be needed.

Virtual interference with CAVs and their environment

The potential for virtual interference with CAVs has formed the basis of many articles in the media. In addition to feeding the CAV incorrect information in order to affect its behaviour, concerns have been expressed about: theft of data, stalking, blackmail, taking control of the vehicle in order to steal it, to kidnap its occupant(s), or to use it as a weapon.

The Computer Misuse Act 1990 (CMA 1990) creates a series of offences of broad application, including: unauthorised access to computer material (Section 1); unauthorised access with intent to commit or facilitate commission of further offences (Section 2); and unauthorised acts with intent to impair the operation of the computer (Section 3). In addition, there is an aggravated offence of unauthorised acts causing, or creating risk of, serious damage, including death or personal injury (Section 3ZA).

As noted by the Law Commission in the 2018 Preliminary Consultation, the drafting of CMA 1990 appears wide enough to deal with

---

[3] The Law Commission's Consultation Paper No. 240 "Automated Vehicles: A Joint Preliminary Discussion Paper", dated 8 November 2018, paragraph 8.14 https://s3-eu-west-2.amazonaws.com/lawcom-prod-storage-11jsxou24uy7q/uploads/2018/11/6.5066_LC_AV-Consultation-Paper-5-November_061118_WEB-1.pdf

the types of conduct outlined above[4]. It will be important, however, to keep the actual behaviour of criminals under review, by taking a pro-active approach to cyber-security and analysing the nature of their attacks, to ensure that CMA 1990 remains fit for purpose.

SUMMARY OF POINTS ON CRIMINALITY

The AEVA 2018 does not deal with any of the potential criminal offences connected with the use of CAVs on our roads
There is already an extensive body of criminal law relating to the use of standard motor vehicles. Many of these offences will be equally applicable to CAVs.
Physical interference with CAVs and their environment is covered to a certain extent by the existing criminal law, but a new "harassment" offence may be required.
Virtual interference with CAVs and their environment is likely to be covered by the Computer Misuse Act 1990, subject to the potential need for amendments or new offences to address novel ways of committing this type of crime.

---

4 The Law Commission's Consultation Paper No. 240 "Automated Vehicles: A Joint Preliminary Discussion Paper", dated 8 November 2018, paragraphs 8.52-8.54 https://s3-eu-west-2.amazonaws.com/lawcom-prod-storage-11jsxou24uy7q/uploads/2018/11/6.5066_LC_AV-Consultation-Paper-5-November_061118_WEB-1.pdf

CHAPTER TWELVE
EMPLOYERS' LIABILITY CLAIMS
SCARLETT MILLIGAN

OBJECTIVES OF THIS CHAPTER

The objectives of this chapter are to:

- Familiarise the reader with employers' current health and safety responsibilities;

- Provide examples of negligence claims which could be brought against employers in respect of CAVs;

- Assess whether, and to what extent, employers may owe a duty of care to those other than their employees for any mishaps caused by CAVs;

- Briefly explain the principles of vicarious liability, and how those principles could apply in a CAV setting; and

- Discuss the role of employment contracts, and whether they could impact upon the aforementioned matters.

INTRODUCTION

There are a number of reasons why employers are likely to be some of the first owners of CAVs, and may also receive a wave of legal claims when things go wrong.

Firstly, there is the extensive testing which CAVs will require prior to being rolled out on public roads. This is likely to be done by (and impact upon) those who are employees, or are in a relationship akin to employment (a significant phrase which we will explore later in this chapter). Whilst those accepting the job of testing CAVs will, to a certain extent, be accepting the inherent risks of the job, what happens if they are injured because of the negligence of their employer or a colleague, for example, in preparing the CAV or the testing area? What if such tests – particularly those on public roads – injure other road users, or damage their cars? Unfortunately, open-road testing in the United States has shown us that there is a distinct possibility of such incidents occurring.

Secondly, it seems probable that the initial CAVs will only be capable and/or permitted to operate on a limited class of roads. In terms of public roads, this is likely to be motorways and other main roads (perhaps portions of the primary route network), all of which are key for employers with a distribution element to their business. These roads are, by and large, inherently easier for CAVs to navigate. They have distinct lanes, minimal changes to direction, and they provide a centralised area to roll out improved connectivity (this includes 5G and the network advances thereafter, as well as CAV-specific facilities such as: CAV-only platooning lanes, Vehicle to Vehicle (V2V) communications, and Vehicle to Infrastructure (V2I) communications). In addition to public roads, CAVs are likely to be well suited to private roads created to connect sites and convey goods; for example, a CAV-specific track designed to carry raw materials between factories and warehouses on a large industrial site.

Thirdly, employers (particularly large employers) are more likely to have the financial capability and incentives to adopt the use of CAVs earlier than most consumers. In addition to having the funds to purchase CAVs, their use of CAVs could also generate significant savings. For example, a CAV-only track connecting an employer's sites and stores would – once up and running with Level 4 and/or Level 5 vehicles – enable employers to save the cost of having human employees drive these vehicles. Even if employees needed to accompany CAVs (perhaps to assist with unloading them, or for security reasons), their time could be spent generating additional profits by undertaking tasks within the

CAV, rather than spending that time driving. Over time, such savings could pay for the initial cost of the CAV itself. If one adds in the efficiency of the CAV not falling ill, not going on holiday, and (supposedly) not making as many mistakes as human drivers, many businesses will see this opportunity as a done deal.

But when CAVs do cause problems – for employees or for third parties – will the employer be held accountable for adopting them in their enterprise and, if so, how and to what extent? This chapter seeks to explain the main legal avenues for redress against an employer, and to provide examples of the facts and legal principles which could be relied upon.

The law encompassing the responsibilities of employers and employees, as well as health and safety standards in the workplace, is vast and worthy of a large tome in and of itself. Necessarily, this chapter only highlights some areas of the law which we consider to be of importance to CAVs. Furthermore, whilst we only refer to domestic statutes and case law, this is an area where additional sources are very important, particularly EU law, HSE Approved Codes of Practice, and other industrial publications. As such, this chapter is by no means a complete guide to this area of law, or a complete summary of the type of claims which may be brought against employers, and should not be taken as such.

THE RELEVANCE OF HEALTH AND SAFETY STATUTES TO A CAV FUTURE

Historically, protections for employees at work developed through the common law. In the latter half of the 20th century, the statute book was expanded significantly to codify specific protections for employees. These developments often stemmed from law-making at the European Union level.

The position since 1 October 2013 (and therefore for this book's purpose) is that health and safety legislation, in and of itself, will not give rise to a civil action for breach of a statutory duty[1]. Breach of these legislative duties may, however, be a criminal offence and pursued by the authorities accordingly, but a detailed explanation of such machinations is beyond this text.

The common view held by practitioners is that this legislation will still have an important function: it will inform a court's understanding of an employer's common law duty of care[2]. In light of this, the approach taken in this chapter is to deal with issues thematically, and to consider the relevant legislation alongside the common law duty of care.

Before we turn to that analysis, it is important to acknowledge that at times there may be a clash between the common law standard of care (reasonableness) and the stipulations of health and safety regulations. For example, the latter may stipulate that employers "shall" take specific actions, or that "practicable" precautions must be put in place. In our view, these tests need not be pitted against one another to decide which test will apply; instead, they are likely to converge, with the language of the regulations informing what it means to take 'reasonable care' in any given context. For example, where regulations stipulate that steps "shall" be taken, these steps are likely to be the court's starting point when considering how a reasonable employer would have discharged their duty of care. In this vein, we note the remarks of Lord Mance in *Baker v Quantum Clothing Limited*[3]:

---

1 Section 69 of the Enterprise and Regulatory Reform Act 2013 amended Section 46 of the Health and Safety at Work Act 1974 to provide that a breach of health and safety regulations *"shall not be actionable except to the extent that regulations under this section so provide"*. All of the regulations cited in this chapter do not give rise to a standalone civil action, unless otherwise specified. One driving factor behind this amendment was the perception of a 'compensation culture' in England and Wales, and a view that employers were unfairly required to pay employees compensation for technical breaches of health and safety legislation, in what were otherwise reasonable working environments which did not pose a risk to employees.

2 See comments to this effect in the Supreme Court case of *Kennedy v Cordia (Services) LLP* [2016] UKSC 6.

3 [2011] UKSC 17, at paragraph 82

> *"The criteria relevant to reasonable practicability must on any view very largely reflect the criteria relevant to satisfaction of the common law duty to take care. Both require consideration of the nature, gravity and imminence of the risk and its consequences, as well as of the nature and proportionality of the steps by which it might be addressed, and a balancing of the one against the other."*

This approach would also seem to align with the possibility of criminal sanctions: if falling below the standards laid down by the legislation could constitute a criminal offence, it would be illogical for this to nonetheless be viewed by the civil courts as 'reasonable', absent justifying circumstances or facts.

NEGLIGENCE CLAIMS AGAINST EMPLOYERS

General Points

Negligence as a cause of action, and what claimants are required to prove on the balance of probabilities, have been discussed at length throughout this book, and will not be repeated here[4]. There are, however, some distinguishing aspects of negligence claims in an employment context which warrant some brief discussion.

The **first** is the breadth of the duty of care, which is to protect the health and safety of an employee in the course of their employment. This duty is not constrained to risks stemming from the specific activities an employee must undertake to fulfil their role, but also includes risks which are incidental to one's employment duties.

In addition, an employer's duty of care consists not only of a duty *not* to take actions which may harm a claimant, but also encompasses the obligation to take *positive* steps to protect employees' health and safety. Thus the duty of care owed by an employer to an employee falls foul of

---

4   See Chapter 1 in particular

the usual 'omissions' rule, i.e. that a duty of care cannot require one to take positive steps to protect someone from risks posed by others.

The **second** is the non-delegable nature of the duty: whilst employers can delegate the *performance* of their duty of care to others, they will nonetheless be held responsible in law if that duty of care is breached[5].

The **third** is that employers may be judged by reference to others' knowledge. Where there is a dispute about an employer's state of knowledge at the time of an accident, the courts will assess this objectively by seeking to establish what was known in the industry at that time (or perhaps published in official or industrial guidance), and an employer will be deemed to have held this knowledge at the material time. Conversely, if the evidence demonstrates that the employer in fact had *"greater than average"* knowledge, the employer will be expected to act based upon what was known to *them*, rather than the knowledge of the industry[6].

To give a relevant example, the Government's latest code of practice on automated vehicle trialling[7] highlights risks such as the loss of contact with a CAV when conducting remote trials, or the fatigue and loss of attention of those monitoring Level 3 CAVs. In line with the principle set out above, employers conducting CAV testing would now be deemed to be aware of these risks. Equally, if a particular employer discovered that remote signals could pose health risks to users, it would need to act accordingly, rather than waiting for this risk to be discovered or acknowledged by the rest of the industry.

In addition, common practice in an industry may be cogent evidence of reasonable behaviour on the part of an employer, though it will not be

---

5 *Wilsons and Clyde Coal Company Limited v English* [1938] A.C. 57. See in particular commentary found on pages 64-65 and 70

6 See *Baker v Quantum Clothing*, footnote 3

7 The Centre for Connected and Autonomous Vehicles' 'Code of Practice: Automated vehicle trialling' dated February 2019: https://assets.publishing.service.gov.uk/government/uploads/system/uploads/attachment_data/file/776511/code-of-practice-automated-vehicle-trialling.pdf

decisive[8]. Again, this does not equate to an entitlement to do nothing until the last possible minute: in *Thompson v Smiths Shiprepairers (North Shields) Limited*[9], the High Court held that once information about risks (and possible protections) became readily available, an employer had a duty to seek out this knowledge, and consider whether it ought to adopt any newfound protections available. The size of an employer, and the resources available to it, will be considerable factors in the 'reasonableness' assessment.

The **fourth** (and final) preliminary point is to acknowledge that employers will not breach their duty by failing to take every possible step open to them. The standard of care, after all, is to do that which a *reasonable* employer would do. The courts have interpreted this duty as requiring an employer to *"…take into account… the likelihood of an accident happening and the gravity of the consequences"*[10], and to balance that risk against their resources and the cost of taking preventative action[11].

Who Will be Owed a Duty of Care by an Employer?

The historic or 'traditional' view was that such an onerous duty of care would only be owed to an employer's employee; others, such as independent contractors, were not owed such a duty of care[12] (though they would be owed an 'ordinary' duty of care, as is owed between persons generally).

---

8 See, for example, *Brown v Rolls Royce Limited* [1960] 1 W.L.R. 210 at page 216-217.

9 [1984] Q.B. 405, particularly at page 422-423

10 *Paris v Stepney Borough Council* [1951] A.C. 367

11 See *Latimer v A.E.C.* [1953] A.C. 643

12 See, for example, judgments such as those in *Inglefield v Macey* (1967) 2 K.I.R. 146 and subsequently *Jones v Minton Construction* (1973) 15 K.I.R. 309, both of which use the archaic language of the liability of a master to his servant, reflecting the way in which employment relationships were historically viewed.

In parallel with society's view of employment, and of common working arrangements, that view has evolved. In summary, the modern approach to this question has been to determine whether contractors were effectively in the same shoes as employees, and to assess who had control of the activities in question.

One of the earliest cases to take this approach was *Denham v Midland Employers Mutual Assurance LD*[13]. The Court of Appeal, in finding that a duty of care was owed to a temporary *'transferred'* employee, focused on the ability of the 'lending' employer to dictate the individual's work, and how it was to be done. The court noted that an employer's duty of care is less likely to arise *"when a man is lent with a machine"* and the parties expect the individual to retain control over the job at hand. This, the court said, was in contrast to the lending of an *"unskilled"* man.

Subsequently, the Court of Appeal in *Lane v Shire Roofing Company (Oxford) Limited*[14] highlighted a number of factors which were relevant to its decision on whether an 'employment' duty of care arose:

- the element of control over the work;

- the business for which the work was being undertaken (i.e. was the independent contractor working for their own business, or for that of another); and

- who had responsibility for the overall safety of the operation and all those involved.

The court went on to draw a distinction between those engaged *"…on 'the lump' to do labouring work (where the men are clearly employees, whatever their tax status may be)"* and a *"specialist sub-contractor"*.

---

13 [1955] 2 Q.B. 437, particularly at page 444

14 [1995] P.I.Q.R. P417, particularly pages 423-424

The Court of Appeal in *Gray v Fire Alarm Fabrication Services Limited*[15] held that an independent contractor could owe a duty of care to a sub-contractor in *"special circumstances"*. In that case, the circumstances justifying the imposition of such a duty were highly fact-specific; they included assertions in the contract (and to health and safety inspectors) that the main contractor would:

- be supervising the work;

- receive a financial premium for this supervision;

- require method statements from the sub-contractor;

- closely liaise with the sub-contractor;

- require the sub-contractor to carry out work to their satisfaction; and

- require the sub-contractor to conform with their directions.

Whilst the general shift in the common law has been toward increasing the class of people to whom an employer may owe a duty of care, it is unclear whether civil courts will interpret health and safety legislation in this way. For example, where specific regulations refer to steps an *'employer'* is required to make, there is a possibility that the courts could rule that such steps are not required in respect of independent contractors. This seems unlikely given that it would return the law to a state of divergence as between employees and independent contractors, but there is certainly an argument to be had about the legislature's intentions.

For the remainder of this chapter, we will refer to 'employers' and 'employees', intending to include all those whom a court would consider to hold (or be owed) the broader 'employment' duty of care,

---

15 [2006] EWCA Civ 1496, see in particular paragraphs 14 and 34

irrespective of whether they are contractually defined as employers or employees.

CAV-related remarks

The question of who will be owed an 'employment' duty of care has been explored at some length because it is an important one in the context of CAVs: individuals involved in CAV testing are more likely to be specialist independent contractors rather than employees. Indeed, a narrower duty of care may be a decisive factor in an employer's decision to utilise independent contractors for testing activities.

Casting our eye further down the line, employers across a range of industries are likely to utilise CAVs, which will lead to further opportunities for independent contractors to be exposed to the problems posed by CAVs.

For the avoidance of doubt, employees or contractors conducting CAV testing will not be able to claim in negligence for *any* injury or damage sustained during a testing exercise. A successful claim would require them to establish that their employer had breached its duty of care; one cannot simply claim for the inherent risks of one's job. Alternatively, those involved in CAV testing may have recourse in a product liability claim against the manufacturer of the CAV (on which see Chapter 9 - "Product Liability Claims").

Thematic Analysis

The numerous employers' duties explored below are perhaps better described as subsets of an employers' overarching duty of care to protect the health and safety of their employees in the course of their employment. The keystone piece of primary legislation in this area, the Health and Safety at Work Act 1974 ("HSWA 1974"), phrases the overarching duty in the following way:

"*s.2. General duties of employers to their employees.*

(1) It shall be the duty of every employer to ensure, so far as is reasonably practicable, the health, safety and welfare at work of all his employees.

The duty to risk assess and provide safe systems of work

These duties are sometimes viewed as 'catch all' or 'fallback' options, but this is to undermine and mis-understand them. It is more apt to see them as the first steps toward fulfilling the overarching duty of care placed on employers.

To provide a safe system of work, an employer ought to consider all of the specific risks and duties engaged by their operations, and target them accordingly. In that vein, a risk assessment can be a central tenet of this duty.

Regulation 3 of the Management of Health and Safety at Work Regulations 1999 ("the Management Regulations") specifies that:

"(3)(1) Every employer shall make a suitable and sufficient assessment of—

(a) the risks to the health and safety of his employees to which they are exposed whilst they are at work; and

(b) the risks to the health and safety of persons not in his employment arising out of or in connection with the conduct by him of his undertaking,

for the purpose of identifying the measures he needs to take to comply with the requirements and prohibitions imposed upon him by or under the relevant statutory provisions"

The appellate courts have echoed the importance of prioritising risk assessments: the recent Supreme Court judgment of *Kennedy v Cordia*

(Services) LLP[16] emphasised that employers' duties and obligations cannot exist in a vacuum, and that risk assessments should be the first step toward understanding the reasonable precautions required in their enterprise.

Conducting this exercise at an early stage effectively forces an employer to engage with the issues of likelihood of risk, costs and practicalities. As it was succinctly put by Lady Justice Smith in *Allison v London Underground Limited*[17]:

> "*What the employer **ought** to have known will be what he **would** have known if he had carried out a suitable and sufficient risk assessment*" (emphasis added)

The suitability and sufficiency of any risk assessment will need to be judged on the facts of the case[18], and will necessarily turn on the nature of the workplace and business operation. As a starting point, the specific themes discussed in the remainder of this section are all likely to be (or should be) key considerations in any risk assessment.

In addition to mandating risk assessments, the Management Regulations require employers to make and give effect "*to such arrangements as are appropriate… for the effective planning, organisation, control, monitoring and review of the preventative and protective measures*"[19], a duty which spans the entire time frame of an employer's operations.

Given the new and uncertain nature of CAV technology, along with its inherent dangers, pre-planning and risk assessments are likely to be crucial for employers who begin to use them in their workplace. Simply assuming that CAVs will function without teething problems would be a dereliction of duty. This is particularly true of Level 3 CAVs, which

---

16 [2016] UKSC 6

17 [2008] EWCA Civ 71, at paragraph 57

18 See *Robert Lee Uren v Corporate Leisure (UK) Limited; Ministry of Defence* [2011] EWCA Civ 66, particularly at paragraph 72

19 Regulation 5(1)

present the additional problems of handovers, intervention, and monitoring fatigue (all discussed in more detail in Chapter 1).

In principle this appears to be a straight-forward responsibility, but the difficulty for employers will lie in attempting to predict how CAVs might malfunction, and identifying appropriate safety mechanisms. For example, while it is entirely foreseeable that a CAV may make a poor driving decision, the precise machinations of that decision (and how a resulting collision may unfold) may not be foreseeable. When considering allegations of negligence, the courts will analyse foreseeability from both a subjective and objective angle: what *did* the employer foresee, and what *should* a reasonable employer have foreseen.

Once a suitable and sufficient risk assessment has been undertaken, an employer's duty will not come to an end. The pace of CAV-related learning will be rapid, and employers will need to make sure they are regularly revising their plans and risk assessments to reflect the known risks of CAVs.

<u>The duty to provide safe equipment</u>

Unlike other aspects of employers' health and safety duties, the duty to provide safe equipment is enshrined in actionable primary legislation. Section 1 of the Employer's Liability (Defective Equipment) Act 1969 ("the Defective Equipment Act 1969") provides:

> "*1(1) Where after the commencement of this Act—*
>
> *(a) an employee suffers personal injury in the course of his employment <u>in consequence of a defect in equipment provided by his employer</u> for the purposes of the employer's business; and*
>
> *(b) the defect is attributable wholly or partly to the fault of a third party (whether identified or not),*
>
> *<u>the injury shall be deemed to be also attributable to negligence on the part of the employer</u> (whether or not he is liable*

in respect of the injury apart from this subsection), but without prejudice to the law relating to contributory negligence and to any remedy by way of contribution or in contract or otherwise which is available to the employer in respect of the injury."

(emphasis added)

Section 1(3) goes on to define *'equipment'* as including vehicles, and it can therefore be assumed that this would apply to CAVs.

The Defective Equipment Act 1969 gives rise to an action in negligence, rather than one for breach of statutory duty. It has the effect of creating a shortcut within the negligence action: provided a defect in the work equipment caused an employee's injury, the employer will be deemed negligent, without the employee needing to establish fault and breach of the duty of care. Indeed, the act is explicit that the defect may be wholly attributable to the fault of another. Whilst this may initially strike one as unjust, it functions as a convenient model which enables vulnerable people to be swiftly compensated by those who directly benefit from taking risks, and who have deeper pockets and insurance protection[20]. Furthermore, employers are likely to have favourable prospects in bringing subsequent claims (or contribution claims) against CAV manufacturers in respect of any defects, with the costs of such actions being met by their insurance coverage.

By contrast, where an employee alleges (without reference to the Defective Equipment Act 1969) that an employer was negligent in failing to identify a defect, then provided that employer can demonstrate that it had undertaken reasonable checks, and had reasonable maintenance and inspection systems in place, a failure to detect a non-obvious defect is unlikely to be regarded by the courts as a breach of the duty to take reasonable care[21].

---

[20] See the Employers' Liability (Compulsory Insurance) Act 1969

[21] See, for example, the judgment in *Henderson v Henry E. Jenkins & Sons* [1970] A.C. 282 (page 300 in particular), where the House of Lords would have permitted such an argument to be raised, but for the failure of the employer to adduce

In addition, the Provision and Use of Work Equipment Regulations 1998 ("PUWER") stipulate that:

> "*Every employer shall ensure that work equipment is so constructed or adapted as **to be suitable** for the purpose for which it is used or provided*"[22] (emphasis added)

Suitability is defined as "*suitable in any respect which it is reasonably foreseeable will affect the health or safety of any person*"[23], and work equipment is defined as "*any machinery, appliance, apparatus, tool or installation for use at work (whether exclusively or not)*"[24].

One of the leading authorities on the definition of work equipment, *Spencer-Franks v Kellogg Brown and Root Limited*[25], suggests that the litmus test ought to be whether the equipment forms "*a useful, practical function in relation to the purposes of that undertaking*". CAVs will undoubtedly fall within that definition.

In addition, see the case of *Crane v Premier Prison Services Limited*[26], where it was taken as a given (though it was not in dispute) that a police van met the definition of 'work equipment'. The dispute centred on an employer's failure to provide handholds for employees to use whilst the van was moving. This failure was held to be a breach of the duty of care, given the need for employees to move around whilst the van was in motion. Similar measures will be essential for Level 4 and

---

evidence of the journeys undertaken by the vehicle, which would necessarily dictate the reasonableness of the employer's maintenance and inspection regime. In such cases, the employer's knowledge of the CAV's use, as well as knowledge of problems with the model of CAV (whether within its own business or across the industry) are likely to be pivotal in assessing the reasonableness of their systems

22 Regulation 4(1)

23 Regulation 4(4)(a)

24 Regulation 2

25 [2008] UKHL 46, see paragraphs 51-53 in particular

26 [2001] C.L.Y. 3298

Level 5 CAVs, where employees may also be moving within the CAV to conduct other work-related tasks.

Such adjustments are likely to go to the 'suitability' of the CAV, rather than to the question of it being 'safe'. Disputes surrounding the safety of a CAV may focus on matters such as hardware or software defects (for example, the inability of a CAV's sensors to detect moving equipment in a factory). Where an injured or affected employee alleges that the CAV was not safe, he or she could also consider suing the CAV's manufacturer[27].

The obligations discussed above only apply in respect of equipment which the employer *chooses* to provide; they do not impose a positive duty to provide specific equipment. Many readers may be familiar with the responsibility of employers to provide personal protective equipment ('PPE'). The Personal Protective Equipment at Work Regulations 1992 do not apply (in large) to *"personal protective equipment used for protection while travelling on a road…"*[28] which would appear to eliminate their application to CAVs.

However, the common law duty of care may nonetheless require employers to provide specific protective equipment for those working in or on CAVs. Consider, for example, employees in a Level 4 or Level 5 delivery vehicle, who may be working in the rear of the CAV alongside stacks of crates, boxes, or vials of hazardous materials. Even if an employee was not due to handle such materials, they could all cause serious injuries in the event of a collision.

Similarly, Regulation 12(1) of PUWER provides that employers *"shall take measures"* to ensure that where persons using work equipment may be exposed to a specified hazard, their exposure to the hazard is *"either prevented, or, where that is not reasonably practicable, adequately controlled"*. A specified hazard which relates to the example above is *"any*

---

27 For more on which see Chapter 9 – "Product Liability Claims"

28 Regulation 3(2)(d)

article or substance falling or being ejected from work equipment"[29]. Regulation 12 appears to suggest that, in this kind of scenario, PPE and training would be a bare minimum, and more could be required:

> "(2) The measures required by paragraph (1) shall–
>
> (a) be measures other than the provision of personal protective equipment or of information, instruction, training and supervision, so far as is reasonably practicable; and
>
> (b) include, where appropriate, measures to minimise the effects of the hazard as well as to reduce the likelihood of the hazard occurring."

Regulations 15 and 16 should be briefly noted in the context of CAVs: they require (respectively) that work equipment shall have stop controls and emergency stop controls where necessary. The authors are not aware of these provisions ever having been raised in the context of vehicles, which is perhaps attributable to current vehicles having brakes, and brake failures being dealt with as defective or unsafe work equipment, rather than under these provisions.

It is therefore unclear whether the courts would consider these particular regulations to be applicable to CAVs (especially in light of their qualified language: "*where appropriate*"). One can envisage a forceful argument in respect of any Level 5 CAVs developed without mechanisms to enable Users in Charge to bring a CAV to an emergency halt. Technically, these controls could be otiose in a Level 5 CAV, and could in fact *create* more problems (for example, fraudsters seeking to induce crashes). In spite of that, it seems unlikely that society would, at least currently, find CAVs palatable without such controls.

In addition to ensuring that work equipment is safe and suitable, employers must also ensure that it is "*maintained in an efficient state, in*

---

[29] Regulation 12(3)(a)

efficient working order and in good repair"[30], and inspected regularly where appropriate, for example, if the equipment is "*exposed to conditions causing deterioration*"[31]. In the context of CAVs, these obligations will translate into practical steps such as:

- conducting or arranging regular safety checks of the CAV before it is put to use (the frequency of such checks being dependent on the regularity of the CAV's use, as well as its travel distance and overall mileage);

- servicing the CAV regularly;

- complying with, and installing, all manufacturers' updates; and

- ensuring that the software is sufficiently robust.

A particularly important aspect of CAV inspection and maintenance is likely to be ensuring that, even with time and use, CAV batteries remain fit for purpose and able to deliver the requisite amount of power for employees to make their journeys in CAVs (making allowances for stops at charging stations). A failure to do so could result in CAVs running out of power in dangerous situations, for example, whilst on a motorway. Most readers will be familiar with the frustrating problem of the battery life of their mobile phone or laptop decreasing over time: whilst currently only a frustration, these problems could have life-threatening consequences for CAVs.

Any work equipment inside an employer's CAV will also attract the aforementioned duties and responsibilities, as it would do today. To return to the example of the autonomous delivery vehicle, this would include equipment used to lift the heavy boxes or crates, or to keep them in situ once loaded.

---

30 Regulation 5, PUWER

31 Regulation 6, PUWER

Providing adequate training and instructions lies at the core of this duty, as does ensuring that employees have the requisite skills and capacity for the tasks entrusted to them. Section 2(2)(c) of the HSWA 1974 expressly specifies that an employer's duty of care for the health, safety and welfare of employees includes:

> *"the provision of such information, instruction, training and supervision as is necessary to ensure, so far as is reasonably practicable, the health and safety at work of his employees"*

See also the requirements in Regulation 13 of the Management Regulations:

- to provide adequate health and safety training;

- to repeat and refresh such training when appropriate; and

- to consider an employee's capabilities when entrusting a task to him or her.

In addition to these general requirements, Regulations 8 and 9 of PUWER specifically require employers to provide adequate training and health and safety information in respect of work equipment.

Employees will certainly require comprehensive training and instructions on the correct and safe operation of CAVs. This will be particularly important for Level 3 and Level 4 vehicles, which pose additional complications in the form of handovers, and a possible requirement to intervene in the CAV's driving[32]. Where Level 3 CAVs are to be utilised for particularly long journeys, the employer may also need to consider some form of suitability testing to ensure that employees have a sufficient attention span to monitor the CAV's driving. There have been suggestions of a separate CAV driving licence,

---

32 See Chapter 1, particularly the section entitled 'Bringing Proceedings in a Non-AEVA Case'

specifically designed to cover such issues[33]. If this requirement were imposed, it would also be part of an employer's duty to check that its employees held a licence (and any further licences required, for larger vehicles for example).

Beyond the routine operation of a CAV, it may be difficult for employers to provide clear training and instructions on the possible dangers posed by CAVs, given that the parameters of those dangers are not yet fully known. As mentioned earlier, it will be for the courts to take a view on what was reasonably foreseeable, and how training and instructions ought to have been delivered. As with risk assessments, employers will need to constantly revisit their provision of information, instructions and training in light of advancements in CAV technology and understanding.

Even with comprehensive training and instructions, some supervision and monitoring in the workplace will be required. The requisite level of supervision will, as usual, depend on the circumstances of the case. Case law suggests that the severity of the risks posed, and an employee's prior experience (or lack thereof) will be particular factors in a court's mind[34].

The first employees to utilise CAVs in the discharge of their employment duties are likely to take some time adjusting to their use. Notwithstanding an initial bout of training, they may fall into poor habits as they let go of any initial inhibitions and become more familiar with CAVs. In this sense, supervision and monitoring will be required in order to confirm that training and instructions have been followed,

---

33 The Law Commission's "Analysis of Responses to Law Commission Consultation Paper No 240", dated 19 June 2019, at paragraphs 5.84-5.85: https://s3-eu-west-2.amazonaws.com/lawcom-prod-storage-11jsxou24uy7q/uploads/2019/06/Automated-Vehicles-Analysis-of-Responses.pdf

34 For example, in the case of *Kerry v Carter* [1969] 1 W.L.R. 1372 (particularly at pages 1375-1376), an employer failed to provide training and instructions on the use of a saw to a young apprentice who had indicated that he had used such a saw previously. Although the apprentice was found to bear some contributory negligence, the employer was held to be negligent in failing to enquire further, and in failing to supervise the apprentice "*more closely*" and for a longer period of time, given his age and status as an apprentice.

and that such arrangements are workable. Employees in the testing context will perhaps require additional scrutiny, given the gravity and/or likelihood of the risks they may face.

Employers must have particular regard to employees becoming inattentive or complacent[35], and such reminders – whether in-car or during training sessions – provide good opportunities to guard against the risks associated with such states of mind.

Whilst our discussion thus far has focused on ways in which an employer may be held liable, it is equally important to emphasise the areas for which they will *not* be held responsible. These include folly or pranks in the workplace, or a failure by an experienced (and trained) employee to select the correct tool for the job at hand[36]. A key – if not determinant – factor in the categorisation of an act as a claimant's failure or folly will be the information, instructions and training he or she has (or has not) been given for the task at hand; for example, it may be difficult to argue that a claimant had foolishly intervened in the driving of a Level 3 CAV if his employer had failed to instruct him that he was prohibited from doing so.

The duty to provide a safe workplace

As mentioned at the outset of this chapter, an employer's duty is not only to ensure safety in respect of an employee's specific duties and activities, but also in the course of employment. This includes a duty to ensure the workplace itself is safe.

Currently, workplace vehicles are, by and large, seen as items of work equipment. In the future, there is a possibility that Level 4 and Level 5 CAVs may be categorised as workplaces in and of themselves: if an employee need not drive or monitor a CAV, and is instead undertaking

---

35 See *Smith v National Coal Board* [1967] 1 W.L.R. 871, page 873-874

36 See for example *Richardson v Stephenson Clarke Limited* [1969] 1 W.L.R. 1695, particularly page 1699

work tasks such as stock-checking within the CAV, the description of 'workplace' sounds equally, if not more, apt.

Indeed, there is no requirement that the duty to provide a safe workplace only applies to an employee's 'main' workplace: the duty may be owed in respect of multiple premises, or even premises owned by another. Whilst the reasonable steps required of remote employers are likely to be different, their duties of care nonetheless remain.

The Workplace (Health, Safety and Welfare) Regulations 1992 are unlikely to apply in substance to CAVs, given that swathes of the regulations currently do not apply to taxed vehicles (or vehicles exempt from tax)[37]. Furthermore, these regulations predominantly make provision for matters which have little practical application in the context of CAVs.

It is, however, still possible that the common law duty of care could recognise CAVs as a workplace, and therefore require employers to take reasonable steps to ensure their safety. The existence of such a duty, combined with an employer's strict liability for CAV defects under the Defective Equipment Act 1969[38], may result in employers having duties not dissimilar from the CAV manufacturers, albeit with rights of recovery against those manufacturers.

Causation

As with all other actions in negligence, claimants will need to establish that breaches of their employer's duty of care caused their injuries or loss.

Contributory Negligence

Deductions for contributory negligence also apply in employment settings. The conduct expected of employees is explicitly recognised in

---

[37] Regulation 3(4)

[38] See section entitled 'The duty to provide safe equipment' above

some of the legislation we have already referred to; for example, Section 7 of the HSWA 1974 mandates that employees have a duty to take reasonable care for the health and safety of themselves and others, and that they shall cooperate with their employer. Similarly, Regulation 14 of the Management Regulations places more specific duties on employees, including the use of machinery and equipment in accordance with training and instructions.

Despite these provisions, judges are often very sympathetic toward the acts of employees, and may be slower to make contributory negligence deductions than they may be in other contexts. The courts have also acknowledged that it is an inherent aspect of human nature to become less attentive in the face of repetitive, familiar tasks. They have therefore refused to make contributory negligence deductions solely on this basis, particularly where the situation has arisen by virtue of an employer's breach of duty[39].

The role of contributory negligence in the context of CAVs is likely to be of more relevance to Level 3 and Level 4 vehicles, for reasons explored in Chapter 1 ("Life before the Automated and Electric Vehicles Act 2018") and Chapter 4 ("Contributory Negligence in the AEVA 2018 ") of this book. When it comes to Level 3 CAVs, the chances are that minds will wander, and that eyes will drift to the passing landscape, or perhaps to an employee's ever-mounting pile of e-mails, which he or she would prefer not to deal with at home. It is unlikely that the courts would condone such behaviour, even in the case of employees who spend the majority of their time travelling. However, the 'momentary inadvertence' line of case law suggests that these employees may be forgiven for oversights, such as a failure to spot that a CAV has entered the wrong mode (absent any significant warning or indication from the CAV).

---

[39] For a recent exposition of this viewpoint, and a recognition that the widespread acts of other employees will be a relevant factor in the court's consideration, see the concise judgment of the Court of Appeal in *Lewis Casson v Spotmix Limited* [2017] EWCA Civ 1994

The particular difficulties associated with monitoring a Level 3 CAV are likely to be seen as a hazard that a reasonable employer must guard against, perhaps by:

- prohibiting the presence of distractions (for example, in the form of e-mails, paperwork, or other technology);

- prohibiting employees who are known to lack focus, or who are known to be tired having previously worked long hours in a CAV[40], to continue in the role of User in Charge; or

- programming the CAV to give regular warnings to grab the attention of the User in Charge.

The latter option may, in fact, be a necessity for all Level 3 CAVs if they are to operate safely on our roads.

NEGLIGENCE CLAIMS BROUGHT BY NON-EMPLOYEES

This section is *not* intended to cover those who are in an employment-like position, such as independent contractors, who are discussed in the other sections of this chapter[41].

---

40 On this note, see the case of *Michael Eyres v Atkinsons Kitchens and Bedrooms Limited* [2007] EWCA Civ 365, where an employer was held negligent for permitting their employee to drive for an excessive number of hours without a proper break, resulting in him falling asleep at the wheel and sustaining serious injuries. The court held that the claimant had to bear some personal responsibility as he must have realised he was at risk of falling asleep. That said, the contribution appears to be low: the court recognised that the claimant was in that position by virtue of his employer's negligence, and raised the contribution deduction from 25% for his failure to wear a seatbelt up to 33%. It seems likely that, in the absence of a failure to wear a seatbelt, the contribution would be somewhat more than 8%; the authors would suggest 10-20% depending on the facts of the particular case.

41 See the section of this chapter entitled 'Who Will be Owed a Duty of Care by an Employer?'

Direct Negligence

Outside of the broad employer-employee duty of care, it must be remembered that employers may also be the recipients of 'ordinary' claims for negligence, brought by other third parties. This is likely to be those somehow affected by the actions of employers, or by the operation of their businesses.

In addition to the common law duty of care which would govern such claims, health and safety legislation explicitly recognises the possible impact that employers and their employees may have on third parties, and impose duties and obligations accordingly:

HSWA 1974

s.3 — General duties of employers and self-employed to persons other than their employees:

> "3 (1) It shall be the duty of every employer to conduct his undertaking in such a way as to ensure, so far as is reasonably practicable, *that persons not in his employment who may be affected thereby are not thereby exposed to risks to their health or safety*"

s.7 — General duties of employees at work:

> "7. It shall be the duty of every employee while at work—
>
> (a) to *take reasonable care for the health and safety of himself and of other persons who may be affected by his acts or omissions at work*…"

Management Regulations

Regulation 3 — Risk assessment:

> "3 (1) Every employer shall make a suitable and sufficient assessment of…

(a) ...

(b) the <u>risks to the health and safety of persons not in his employment arising out of or in connection with the conduct by him of his undertaking</u>..."

(emphasis added)

Whilst our focus is on civil claims which may be brought against *employers*, we briefly note that any actions contrary to the legislation above may separately render an *employee* to be in breach of health and safety law, and therefore subject to criminal penalties.

Third parties wishing to bring a civil claim can do so in two principal ways: by bringing a direct allegation of negligence against the employer, or by alleging that the employer is vicariously liable for the negligence of another.

Where a third party alleges that the employer was itself negligent, this can still be through the acts of another; for example, where a managing director (acting with authority) prematurely instructs employees to test-drive Level 3 CAVs, resulting in a collision. In such cases, it is the employer - and its instructions or actions - who is negligent, rather than those further down the chain.

'Indirect' Negligence, a.k.a. Vicarious Liability

An employer's liability for the negligent actions of those 'further down the chain' is known as vicarious liability. There are two key determinants of whether an employer will be found vicariously liable: the relationship between the negligent individual and the employer; and the nature of the negligent act and its connection to the employer.

The first issue: the vicarious relationship

In the historic case of *Various Claimants v Institute of the Brothers of the Christian Schools*[42], the Supreme Court held that for vicarious liability to arise, the relationship need not necessarily be one of employer and employee, but could include a relationship "*akin to an employment relationship*". Whilst the application of this test will turn on the facts of a particular case, it undoubtedly expands the scope of an employer's liability for the acts of others.

The second issue: the nature of the alleged act

Clearly it would not be just nor fair for an employer to be held accountable for *every* negligent act of their employees, irrespective of the context in which they occurred. The vexing question for the courts has been where to draw the line between acts which should be attributed to the principal, and those which should not.

Again, *Various Claimants* offers some insight: vicarious liability will arise where negligent acts were committed "*in the course of employment*". This includes acts required by virtue of the employment relationship, and also where there is a "*close connection*" between the relationship and the negligent act(s). In reviewing the case law surrounding what is said to constitute a "*close connection*", the Supreme Court offered the following explanation in the context of sexual abuse:

> "*Vicarious liability is imposed where a defendant, whose relationship with the abuser put it in a position to use the abuser to carry on its business or to further its own interests, has done so in a manner which __has created or significantly enhanced the risk__ that the victim or victims would suffer the relevant abuse. The essential closeness of connection between the relationship between the defendant and the tortfeasor and the acts of abuse thus involves __a strong causative link… Creation of risk is not enough, of itself, to give rise to__*

---

42 [2012] UKSC 56

vicarious liability for abuse but it is always likely to be an important element in the facts that give rise to such liability."[43]

(emphasis added)

Since *Various Claimants*, the courts have continued to grapple with the question of when a close connection will be made out on the facts. At the time of writing, the Supreme Court's latest commentary on this issue can be found in *Mohamud v WM Morrison Supermarkets PLC*[44]:

*"In the simplest terms, the court has to consider two matters. The first question is what functions or "field of activities" have been entrusted by the employer to the employee, or, in everyday language, **what was the nature of his job**. As has been emphasised in several cases, this question must be addressed broadly...*

*Secondly, the court must decide **whether there was sufficient connection** between the position in which he was employed and his wrongful conduct **to make it right for the employer to be held liable** under the principle of social justice..."*

(emphasis added)

This is unlikely to be the last time that the appellate courts pronounce on this issue, and the authors are aware of at least one vicarious liability appeal heading to the Supreme Court[45].

An employer would be held vicariously liable for an employee's negligent driving, use, or monitoring of a CAV, provided that such actions took place in the context of the employee's ordinary work duties, as such actions are *"in the course of employment"*. The difficulty, of course,

---

43 ibid – see paragraph 62, and paragraphs 86-87

44 [2016] UKSC 11, at paragraphs 44 and 45

45 See the judgment and outstanding appeal in *Barclays Bank Plc v Various Claimants* [2018] EWCA Civ 1670, where the defendant argued that it ought not to be vicariously liable for the actions of an independent contractor, and that the law ought to draw a 'bright' dividing line between employees and independent contractors for the purposes of vicarious liability

comes with ascertaining whether other negligent acts meet the criterion of "*close connection*". Case law (albeit preceding *Various Claimants*) suggests that the following acts could be attributable to employers:

- permitting others to take the role of User in Charge[46];

- enlisting others to assist with duties or tasks[47] (for example, asking another to monitor a Level 3 CAV, despite being prohibited to do so);

- negligent actions somehow associated with the role of User in Charge (for example, negligence whilst using an electricity charging point[48])

On the other hand, acts which are less likely to be attributed to employers include: offering a lift to another where this has been prohibited (and where the person receiving the lift is also aware of this prohibition[49]), or using an employer's CAV for personal use outside of work (again, where this has been prohibited).

---

46 See *Ilkiw v Samuels* [1963] 1 W.L.R. 991, particularly at pages 998-999

47 See *Rose v Plenty* [1976] 1 W.L.R. 141, particularly at pages 144-145 where Lord Justice Denning had the following to say about acts prohibited by employers: "*In considering whether a prohibited act was within the course of the employment, it depends very much on the purpose for which it is done. If it is done for his employers' business, it is usually done in the course of his employment, even though it is a prohibited act.*"

48 Extrapolated from the case of *Century Insurance Company Limited v Northern Ireland Road Transport Board* [1942] A.C. 509, where the House of Lords found an employer to be vicariously liable for its employee, who threw away a lighted match whilst delivering petrol, resulting in an explosion.

49 See, for example, *Twine v Bean's Express Limited* [1946] 1 All E.R. 202, particularly paragraphs 204E-G

CONTRACTUAL CLAIMS AGAINST EMPLOYERS

Little can be said in a book like this about the possibility of an employee relying on his or her contract to bring a contractual claim against his or her employer: this will turn almost entirely on the substance of the contract itself. However, it is worth reminding readers that a claim for breach of contract does not (absent a term specifying the contrary) require a claimant to demonstrate that his or her employer was at fault. This is in direct contrast with the negligence actions discussed above, where fault is the crux of the liability.

In addition, there are two key points to be made about the impact of a contract upon a negligence claim. The first is a reminder that Section 2 of the Unfair Contract Terms Act 1977 prohibits contracts from excluding or restricting negligence-based liability for personal injury or death. The second is to highlight that the specific terms of a contract may themselves determine whether a court finds that the broader 'employment' duty of care existed, or whether a paired-back 'ordinary' duty of care will be owed[50].

THE POSSIBLE FUTURE OF EMPLOYERS LIABILITY CLAIMS IN THE CAV CONTEXT

Whilst this chapter has necessarily focused on many of the ways in which employers could be held liable for CAV mishaps, it is important not to lose sight of the fact that CAVs (particularly Level 5 CAVs) will hopefully reduce the number of accidents and injuries in the workplace dramatically. Their technology will provide them with the tools to

---

50 See, for example, *Arthur White (Contractors) Ltd v Tarmac Civil Engineering Limited* [1967] 1 W.L.R. 1508 (particularly page 1516) in which the contract for hire of a mechanical excavator and its accompanying driver specified that the hirer "*alone*" would be responsible *"for all claims' which arise in connection with the operation of the plant by the driver or operator who, as between the parties, is to be regarded as their servant or agent"*, resulting in the court finding that an employer's duty of care was owed by the hirer to the driver . A more recent example is the case of *Gray v Fire Alarm Fabrication Services Limited:* see the reference (and associated text) at footnote 15

avoid common risks and mishaps; for example, a CAV in a busy delivery warehouse should, by virtue of its numerous sensors, detect emerging humans and vehicles which may have been hidden to even the most diligent human eye. As has become a running theme in this book, it is the interim problems associated with Level 3 (and, to a lesser extent, Level 4) CAVs which may be a major source of liability for employers.

There is, of course, a possibility that future legislation will provide comprehensive specifications as to: how CAVs may be used in an employment context; the safety and behavioural standards expected in such circumstances; and where liability will fall. Until such legislation is enacted (indeed, if it is) the current employers' liability framework is – as has hopefully been demonstrated in this chapter - capable of accommodating claims involving CAVs, albeit perhaps with some minor amendments and alterations.

Finally, it remains to be acknowledged that, as currently drafted, the AEVA 2018 will apply to any accidents which employees experience whilst in their employer's CAV, provided that those accidents meet the AEVA's criteria. These criteria were discussed in more depth in Part Two of this book. As explained in Chapter 2, the AEVA 2018 will enable injured claimants to claim compensation directly from the insurer of the CAV (which may or may not be the employer's insurer). The insurer will then have an opportunity to make a claim against a manufacturer (if the CAV was defective), and/or against the claimant's employer (in the event that it breached its duty to provide safe work equipment, or a safe workplace).

SUMMARY OF POINTS ON CLAIMS AGAINST EMPLOYERS

The duty owed by an employer to its employee is a broad one, and is non-delegable
Independent contractors may also be owed a broad duty of care by those seen as their 'employers'
A key aspect of an employer's duty of care will be organisational pre-planning and risk assessments, which will need to be regularly updated in line with CAV developments
CAVs will be 'work equipment', and employers have duties to ensure such equipment is suitable and free from defects
As part of their obligation to provide competent staff, employers will need to provide adequate information, instructions and training on CAVs, particularly in respect of Level 3 CAVs
Level 4 and Level 5 CAVs may be deemed 'workplaces', thus placing an additional duty on employers to take reasonable care to ensure that CAVs are safe
Third parties may also bring claims in negligence where they are affected by an employer's CAV use
Third parties may bring such claims by alleging negligence directly against the employer, or by alleging that the employer is vicariously liable for the failings of its employee
Depending on their contract, employees may also have a claim against their employer for breach of contract

CHAPTER THIRTEEN
EQUALITY
ALEX GLASSBROOK

"*The ability to travel and to get about is important to all of us. Without it we cannot get to work, do the shopping, visit family and friends or places of entertainment, in short be part of the community. Difficulties with transport are one of the two most common barriers to work for people with impairments. Of the 12m disabled people in the United Kingdom, one tenth, that is 1.2m people, are wheelchair users and more than a quarter of these are under the age of 60 (Papworth Trust, Disability in the United Kingdom 2014, Facts and figures).*"

Lady Hale, dissenting in part (in the minority favouring the award of damages), in *FirstGroup Plc v Paulley* [2017] UKSC 4, paragraph 93.

INTRODUCTION

Connected and Autonomous Vehicles have been praised, as a concept, for providing individual, independent mobility to those currently deemed incapable of driving a car.

That is a powerful promise. Its reality will depend not just upon the technology (and, in particular, progress towards fully automated driving at Level 5) but also upon the ways in which the supporting infrastructure of fully automated driving is designed. That will include road design and infrastructure. The laws of transport will be part of that essential infrastructure.

Mobility is due to be considered by the Law Commissions, in the next part of their AV law consultation. The potential of AV technology to achieve mobility for everyone is undeniable. But it will exist in a world designed for drivers. Universal AV mobility will need strong support, including by rigorous real-world testing, and physical and legal infrastructure.

Equality law in relation to road transport is a developing field. As the judgments in the *Paulley* case show, it is one in which different perceptions of justice and practicality conflict.

The practical considerations in future CAV equality cases are likely to be very different: on one possible version of our future, it could be a world without drivers. Would that work?

OBJECTIVES OF THIS CHAPTER

This chapter:

- Outlines the main, current laws of equality which apply to road transport

- Asks what equality law considerations arise for the movement towards fully automated (Level 5) transport.

This chapter is not intended as a comprehensive guide to the laws of equality in the UK, even in the road transport context (where the material is voluminous). This chapter provides the present context, from which future equality laws relating to CAVs will evolve.

The Equality Laws of Road Transport

The Equality Act 2010 in General Terms

The Equality Act 2010 provides, in general terms:

- In its first section, a public sector duty regarding socio-economic inequalities (we quote that section in full, below).

- Protection against discrimination for 9 characteristics, listed in section 4: age, disability, gender reassignment, marriage and civil partnership, pregnancy and maternity, race, religion or belief, sex, sexual orientation.

- Discrimination is categorised into 3 broad types: direct (less favourable treatment, because of a protected characteristic, than would have been given to others: section 13), combined (less favourable treatment because of a combination of 2 protected characteristics than would have been given to another lacking either characteristic: section 14) and indirect discrimination (the application of a provision, criterion or practice which is generally applied but discriminatory in relation to a protected characteristic in that it puts people with that characteristic, and the particular claimant, at a disadvantage which is disproportionate to the legitimate aim of the provision: section 19).

- There are particular prohibitions of discrimination arising from the protected characteristics of disability (section 15), gender reassignment: absences from work (section 16) and for work and non-work cases for pregnancy and maternity (sections 17 and 18).

- Sections 20 to 22 describe the duty to make reasonable adjustments for disabled persons.

The Equality Act 2010 prescribes, in its first section, a public sector duty regarding socio-economic inequalities. It also provides specifically for equality in transport, in certain situations. Those sections are as follows.

Section 1(1) of the Equality Act sets out the public sector duty regarding socio-economic inequalities:

"An authority to which this section applies must, when making decisions of a strategic nature about how to exercise its functions, have due regard to the desirability of exercising them in a way that is

> *designed to reduce the inequalities of outcome which result from socio-economic disadvantage."*

In Part 3 of the Act ("Services and public functions", which does not apply to protected characteristics of age under 18 nor marriage and civil partnership: section 28(1)), section 29 sets out the prohibitions against discrimination in provision of services etc (we quote only subsections (1) and (2)):

> *"(1) A person (a "service-provider") concerned with the provision of a service to the public or a section of the public (for payment or not) must not discriminate against a person requiring the service by not providing the person with the service.*
>
> *(2) A service-provider (A) must not, in providing the service, discriminate against a person (B)—*
>
> *(a) as to the terms on which A provides the service to B;*
>
> *(b) by terminating the provision of the service to B;*
>
> *(c) by subjecting B to any other detriment."*

And, at section 29(7):

> *"A duty to make reasonable adjustments applies to—*
>
> *(a) a service-provider …;*
>
> *(b) a person who exercises a public function that is not the provision of a service to the public or a section of the public."*

Section 149 ("public sector equality duty") provides as follows:

> *"(1) A public authority must, in the exercise of its functions, have due regard to the need to—*

(a) eliminate discrimination, harassment, victimisation and any other conduct that is prohibited by or under this Act;

(b) advance equality of opportunity between persons who share a relevant protected characteristic and persons who do not share it;

(c) foster good relations between persons who share a relevant protected characteristic and persons who do not share it.

(2) A person who is not a public authority but who exercises public functions must, in the exercise of those functions, have due regard to the matters mentioned in subsection (1).

(3) Having due regard to the need to advance equality of opportunity between persons who share a relevant protected characteristic and persons who do not share it involves having due regard, in particular, to the need to—

(a) remove or minimise disadvantages suffered by persons who share a relevant protected characteristic that are connected to that characteristic;

(b) take steps to meet the needs of persons who share a relevant protected characteristic that are different from the needs of persons who do not share it;

(c) encourage persons who share a relevant protected characteristic to participate in public life or in any other activity in which participation by such persons is disproportionately low.

(4) The steps involved in meeting the needs of disabled persons that are different from the needs of persons who are not disabled include, in particular, steps to take account of disabled persons' disabilities.

(5) Having due regard to the need to foster good relations between persons who share a relevant protected characteristic and persons who

do not share it involves having due regard, in particular, to the need to—

(a) tackle prejudice, and

(b) promote understanding.

(6) Compliance with the duties in this section may involve treating some persons more favourably than others; but that is not to be taken as permitting conduct that would otherwise be prohibited by or under this Act.

(7) The relevant protected characteristics are—

age;

disability;

gender reassignment;

pregnancy and maternity;

race;

religion or belief;

sex;

sexual orientation.

(8) A reference to conduct that is prohibited by or under this Act includes a reference to—

(a) a breach of an equality clause or rule;

(b) a breach of a non-discrimination rule.

(9) Schedule 18 (exceptions) has effect."

(Schedule 18 excepts various bodies and activities which, at present, seem unlikely to affect the equality duty insofar as it might otherwise relate to CAVs).

Section 158 ("positive action") provides (insofar as might be relevant to CAVs) as follows:

"(1) This section applies if a person (P) reasonably thinks that—

(a) persons who share a protected characteristic suffer a disadvantage connected to the characteristic,

(b) persons who share a protected characteristic have needs that are different from the needs of persons who do not share it, or

(c) participation in an activity by persons who share a protected characteristic is disproportionately low.

(2) This Act does not prohibit P from taking any action which is a proportionate means of achieving the aim of—

(a) enabling or encouraging persons who share the protected characteristic to overcome or minimise that disadvantage,

(b) meeting those needs, or

(c) enabling or encouraging persons who share the protected characteristic to participate in that activity.

(3) Regulations may specify action, or descriptions of action, to which subsection (2) does not apply.

…

(6) This section does not enable P to do anything that is prohibited by or under an enactment other than this Act."

The Equality Act 2010 in relation to Road Transport

Specifically, in relation to transport (and to our present context of road transport), the subject of Part 12 of the Equality Act is "Disabled Persons: Transport". Chapter 1 of Part 12 (sections 160 to 173) deals with "Taxis etc" and Chapter 2, including 2A, (sections 174 to 181D) deals with "public service vehicles".

Of those sections, the following are the most relevant:

In Chapter 1 of Part 12 ("Taxis etc"):

- Sections 160 to 164 are not yet entirely in force. These sections seek to secure safe and comfortable travel in taxis for disabled persons, including while using wheelchairs.

- Sections 165 to 167 are in force. These prescribe the assistive duties on the driver of a designated, wheelchair-accessible taxi hired by a disabled person in a wheelchair, or by their companion, including the prohibition against making any additional charge for such assistance, and the listing of designated vehicles.

- Sections 168 to 171 are in force. These provide equivalent duties in the case of a disabled person who is accompanied by their assistance dog. section 170 makes it an offence for a private hire vehicle to refuse to accept a booking from a disabled person on the ground that they will be accompanied by their assistance dog, and by making an additional charge for the same. A private hire driver might seek medical exemption by certificate and appeal the refusal of the same (sections 171, 172).

In Chapters 2 and 2A of Part 12 ("Public service vehicles", all sections of which (174 to 181D) are in force):

- Sections 174 to 180 provide that the Secretary of State may make regulations[1] for securing disabled persons' ability to get on and off regulated public service vehicles "without unreasonable difficulty" and to travel in such vehicles "in safety and reasonable comfort", particularly when using wheelchairs. Compliance must be certified, by inspection of the vehicle (section 176). It is an offence (section 175) to contravene those regulations.

- Sections 181A to 181D empower the Secretary of State to make regulations requiring operators of bus services to make available to disabled passengers information relating to their service. The Secretary of State must issue guidance about the duties imposed on operators by regulations under section 181A (section 181C).

The Paulley case

In *FirstGroup Plc v Paulley* [2017] UKSC 4, the Supreme Court considered "the lawfulness of a bus company's policy in relation to the use of the space provided for wheelchair users on its buses"

On 24 February 2012, Mr Paulley, a wheelchair user:

"… arrived at Wetherby bus station, expecting to catch the 9.40 bus ("the Bus") to Leeds. On arrival at Leeds he intended to catch the train to Stalybridge to meet his parents for lunch. The Bus was operated by a subsidiary of FirstGroup PLC ("FirstGroup"), which is the parent company of a group of companies which operates a total of about 6,300 buses. The Bus was equipped with a lowering platform

---

1 See the Public Service Vehicles Accessibility Regulations 2000/1970. But see also Lady Hale's partially-dissenting judgment in *FirstGroup Plc v Paulley* [2019] UKSC 6 at paragraphs 95, 96 and at 100: "… mere compliance with the earlier regulations, both as to the provision of the wheelchair space and affording access to it, is not necessarily enough. Parliament must have contemplated, in passing the 2005 [*Disability Discrimination*] Act [*note: the precursor to the 2010 Act*] that other adjustments to "business as usual" might be needed in order to reduce the difficulties faced by disabled people in using public transport services" [100]. Lord Clarke concurred [157].

and a wheelchair ramp. The Bus also had a space (a "space") for wheelchairs, which included a sign that read "Please give up this space if needed for a wheelchair user."

"When Mr Paulley started to board the Bus, the driver, Mr Britcliffe, asked him to wait because the space was occupied by a woman with a sleeping child in a pushchair. The space had a sign with the familiar designation of a wheelchair sign, and in addition it had a notice ("the Notice") saying "Please give up this space for a wheelchair user". Mr Britcliffe asked the woman to fold down her pushchair and move out of the space so that Mr Paulley could occupy it in his wheelchair. She replied that her pushchair did not fold down, and refused to move. Mr Paulley then asked whether he could fold down his wheelchair and use an ordinary passenger seat. Mr Britcliffe refused that request, because there was no safe way of securing the wheelchair and the Bus had to take a rather winding route."

"As a result, Mr Paulley had to wait for the next bus, which left around 20 minutes later. The consequence of this was that Mr Paulley missed his train at Leeds, and had to take a later train which arrived at Stalybridge an hour later than he had planned."[2]

The bus company's policy, at the time of the incident, was quoted in paragraph 7 of the judgment, as follows:

"As part of our commitment to providing accessible travel for wheelchair users virtually all our buses have a dedicated area for wheelchair users; other passengers are asked to give up the space for wheelchairs. ... If the bus is full or if there is already a wheelchair user on board unfortunately we will not be able to carry another wheelchair user. ... Wheelchairs do not have priority over buggies, but to ensure that all our customers are treated fairly and with consideration, other customers are asked to move to another part of the

---

2 Paragraphs 2 to 4 of the judgment of the Supreme Court in *Paulley*.

bus to allow you to board. Unfortunately, if a fellow passenger refuses to move you will need to wait for the next bus."

As the Supreme Court noted, that policy had "changed somewhat" by the time the case came to trial, to the following:

> "*As part of our commitment to providing accessible travel for wheelchair users virtually all our buses have a dedicated wheelchair area for wheelchair users; other passengers are asked to give up the space for wheelchairs. ...*
>
> *Wheelchair users have priority use of the wheelchair space. If this is occupied with a buggy, standing passengers or otherwise full, and there is space elsewhere on the vehicle, the driver will ask that it is made free for a wheelchair user. Please note that the driver has no power to compel passengers to move in this way and is reliant on the goodwill of the passengers concerned. Unfortunately, if a fellow passenger refuses to move you will need to wait for the next bus.*"

Mr Paulley sought damages for breach of section 29(2) of the Equality Act 2010, on the ground that FirstGroup had failed to make "reasonable adjustments" to its policy.

The Recorder allowed Mr Paulley's claim. The Court of Appeal reversed that decision, and found against him. He appealed to the Supreme Court, who restored the Recorder's decision that the bus company had acted unlawfully, though (by a bare majority of 4 to 3 Justices) disallowing the Recorder's award of damages to Mr Paulley.

The majority's disallowance of damages was on the ground that the Recorder had not considered whether the reasonable adjustment (which, it was agreed, did not require the bus company to go so far as physical ejecting the uncooperative passenger) would have resulted in any different outcome. The minority concluded that there was at least a real prospect that the suggested adjustment (a firm request by the driver of the passenger refusing to move, followed if necessary by firmer action such as stopping the bus) would have resulted in a different outcome, so as to merit damages.

Discussion

The judgments of the Supreme Court in *Paulley* demonstrate a variety of perceptions as to the practicalities of transport for disabled persons. There was disagreement between the Justices, for example, as to the meaning of the judgment below.

There was also a difference in judicial perception as to whether bad manners should be justiciable by the courts (Lord Sumption finding not, in the absence of an alternatively drafted regulation – see paragraphs 88 to 92 – though not considering this to be a case appropriate for a dissent; Lady Hale considering the question of "*how a priority policy might be enforced against recalcitrant passengers*" to be "*something of a red herring*", based both upon her experience of the London underground and the clarity of the law in question – see paragraphs 106 to 109).

EQUALITY AND FULLY DRIVERLESS ROAD TRANSPORT (LEVEL 5)

The *Paulley* case involved a driver, upon whose reactions the case turned.

Will future public transport be fully driverless? Even if the technology permits it, will it be practically desirable?

Those are not legal questions, though the law must operate on the basis of practicality (as both judicial camps in the Supreme Court in *Paulley* plainly assumed, albeit with different results on the question of damages).

The Human Factor

The "human factor" has already been discussed, in the context of the Level 3 or 4 CAV driver's duty (if it is a duty) to intervene in the automated driving of the vehicle, should the circumstances require.

In the different context of equality laws applied to public transport, the same phrase arises but in a different way.

The practicality of entirely automating a sensitive situation has been questioned by the courts already: for example in the case of *Bridges*[3], discussed in the "Data and Privacy" chapter, where the Divisional Court emphasised the presence of human supervision as evidence of compliance with data protection law and practice.

Will technology provide every user with the assurance of safety? This debate has been prominent in all discussions of driverless cars, but only from the perspective of a human qualified to drive, who feels a lack of confidence in the driving skill of a machine.

The same issue has not been considered widely, or at all, from the perspective of a user of an AV who feels a lack of confidence because they are alone, in a machine driven by itself, and do not have immediate access to human assistance. This point arises more readily when AV law is considered from the Equality perspective, perhaps especially from the perspective of a disabled person. Concerns in that situation might include security.

Might, paradoxically, a CAV providing public transport require a human presence: not for human driving skills but for human skills of diplomacy, in a situation of human conflict?

---

3 *The Queen (on the application of Edward Bridges) v The Chief Constable of South Wales Police (Defendant), the Secretary of State for the Home Department (Interested party), the Information Commissioner and the Surveillance Camera Commissioner (Interveners)* [2019] EWHC 2341 (Admin), 4 September 2019)

CONCLUSION

Connected technologies in vehicles and road infrastructure, and carried by road users, will be essential features of future vehicles. In the field of personal transport, CAV technology will, if achieved to Levels 4 and 5, revolutionise personal independence for those currently unable to qualify to drive. Equality law should continue to progress and adapt to that future.

The Equality laws as they affect public road transport, even in a world of conventional vehicles, have progressed but, as the judgments in *Paulley* show, they have yet to settle.

Looking into the future of Connected and Autonomous Vehicles through the lens of Equality law raises a significant, question for future public transport: can safe and secure public transport, available to all, be driverless transport? Does it instead need the human factor? Do we need more tram conductors?

SUMMARY OF POINTS ON EQUALITY

CAVs should be of high benefit to those presently unable to qualify to drive. Equality law should adapt to enable that technology.
Equality laws affecting conventional road transport have been slow to develop but have increased in specificity
There is still not a common approach to the practical questions, upon which the interpretation of Equality laws affecting transport rest
Human intervention seems essential to ensuring safe, secure and universally available public transport
Paradoxically, will the driverless revolution in personal transport emphasise the need for human supervision in public transport?

MORE BOOKS BY LAW BRIEF PUBLISHING

A selection of our other titles available now:-

'Ellis on Credit Hire – Sixth Edition' by Aidan Ellis & Tim Kevan
'Tackling Disclosure in the Criminal Courts – A Practitioner's Guide' by Narita Bahra QC & Don Ramble
'A Practical Guide to TOLATA Claims' by Greg Williams
'Artificial Intelligence – The Practical Legal Issues' by John Buyers
'A Practical Guide to Prison Injury Claims' by Malcolm Johnson
'A Practical Guide to Hackney Carriage Licensing in London' by Stuart Jessop
'A Practical Guide to Advising Clients at the Police Station' by Colin Stephen McKeown-Beaumont
'A Practical Guide to Antisocial Behaviour Injunctions' by Iain Wightwick
'Practical Mediation: A Guide for Mediators, Advocates, Advisers, Lawyers, and Students in Civil, Commercial, Business, Property, Workplace, and Employment Cases' by Jonathan Dingle with John Sephton
'Planning Obligations Demystified: A Practical Guide to Planning Obligations and Section 106 Agreements' by Bob Mc Geady & Meyric Lewis
'A Practical Guide to Crofting Law' by Brian Inkster
'A Practical Guide to Spousal Maintenance' by Liz Cowell
'A Practical Guide to the Law of Domain Names and Cybersquatting' by Andrew Clemson
'A Practical Guide to the Law of Gender Pay Gap Reporting' by Harini Iyengar
'A Practical Guide to the Rights of Grandparents in Children Proceedings' by Stuart Barlow
'NHS Whistleblowing and the Law' by Joseph England
'Employment Law and the Gig Economy' by Nigel Mackay & Annie Powell
'A Practical Guide to the General Data Protection Regulation (GDPR)' by Keith Markham

'A Practical Guide to Noise Induced Hearing Loss (NIHL) Claims' by Andrew Mckie, Ian Skeate, Gareth McAloon
'An Introduction to Beauty Negligence Claims – A Practical Guide for the Personal Injury Practitioner' by Greg Almond
'Intercompany Agreements for Transfer Pricing Compliance' by Paul Sutton
'Zen and the Art of Mediation' by Martin Plowman
'A Practical Guide to the SRA Principles, Individual and Law Firm Codes of Conduct 2019 – What Every Law Firm Needs to Know' by Paul Bennett
'A Practical Guide to Licensing Law for Commercial Property Lawyers' by Niall McCann & Richard Williams
'A Practical Guide to Adoption for Family Lawyers' by Graham Pegg
'Essential Motor Finance Law for the Busy Practitioner' by Richard Humphreys
'A Practical Guide to Industrial Disease Claims' by Andrew Mckie & Ian Skeate
'A Practical Guide to the Law of Armed Conflict' by Jo Morris & Libby Anderson
'A Practical Guide to Redundancy' by Philip Hyland
'A Practical Guide to Vicarious Liability' by Mariel Irvine
'A Practical Guide to Claims Arising from Delays in Diagnosing Cancer' by Bella Webb
'A Practical Guide to Applications for Landlord's Consent and Variation of Leases' by Mark Shelton
'A Practical Guide to Relief from Sanctions Post-Mitchell and Denton' by Peter Causton
'Butler's Equine Tax Planning: 2nd Edition' by Julie Butler
'A Practical Guide to Equity Release for Advisors' by Paul Sams
'A Practical Guide to Unlawful Eviction and Harassment' by Stephanie Lovegrove
'A Practical Guide to the Law Relating to Food' by Ian Thomas
'A Practical Guide to the Ending of Assured Shorthold Tenancies' by Elizabeth Dwomoh
'A Practical Guide to Financial Services Claims' by Chris Hegarty
'The Law of Houses in Multiple Occupation: A Practical Guide to HMO Proceedings' by Julian Hunt
'A Practical Guide to Unlawful Eviction and Harassment' by Stephanie Lovegrove

'A Practical Guide to Solicitor and Client Costs' by Robin Dunne
'A Practical Guide to Wrongful Conception, Wrongful Birth and Wrongful Life Claims' by Rebecca Greenstreet
'Occupiers, Highways and Defective Premises Claims: A Practical Guide Post-Jackson – 2nd Edition' by Andrew Mckie
'A Practical Guide to Financial Ombudsman Service Claims' by Adam Temple & Robert Scrivenor
'A Practical Guide to the Law of Enfranchisement and Lease Extension' by Paul Sams
'A Practical Guide to Marketing for Lawyers – 2nd Edition' by Catherine Bailey & Jennet Ingram
'A Practical Guide to Advising Schools on Employment Law' by Jonathan Holden
'Certificates of Lawful Use and Development: A Guide to Making and Determining Applications' by Bob Mc Geady & Meyric Lewis
'A Practical Guide to the Law of Dilapidations' by Mark Shelton
'A Guide to Consent in Clinical Negligence Post-Montgomery' by Lauren Sutherland QC
'A Practical Guide to Running Housing Disrepair and Cavity Wall Claims: 2nd Edition' by Andrew Mckie & Ian Skeate
'A Practical Guide to Digital and Social Media Law for Lawyers' by Sherree Westell
'A Practical Guide to Holiday Sickness Claims – 2nd Edition' by Andrew Mckie & Ian Skeate
'A Practical Guide to Elderly Law' by Justin Patten
'Arguments and Tactics for Personal Injury and Clinical Negligence Claims' by Dorian Williams
'A Practical Guide to QOCS and Fundamental Dishonesty' by James Bentley
'A Practical Guide to Drone Law' by Rufus Ballaster, Andrew Firman, Eleanor Clot
'A Practical Guide to Compliance for Personal Injury Firms Working With Claims Management Companies' by Paul Bennett
'A Practical Guide to the Landlord and Tenant Act 1954: Commercial Tenancies' by Richard Hayes & David Sawtell
'A Practical Guide to Psychiatric Claims in Personal Injury' by Liam Ryan
'A Practical Guide to Dog Law for Owners and Others' by Andrea Pitt

'RTA Allegations of Fraud in a Post-Jackson Era: The Handbook – 2nd Edition' by Andrew Mckie
'RTA Personal Injury Claims: A Practical Guide Post-Jackson' by Andrew Mckie
'On Experts: CPR35 for Lawyers and Experts' by David Boyle
'An Introduction to Personal Injury Law' by David Boyle
'A Practical Guide to Claims Arising From Accidents Abroad and Travel Claims' by Andrew Mckie & Ian Skeate
'A Practical Guide to Chronic Pain Claims' by Pankaj Madan
'A Practical Guide to Claims Arising from Fatal Accidents' by James Patience
'A Practical Approach to Clinical Negligence Post-Jackson' by Geoffrey Simpson-Scott
'Employers' Liability Claims: A Practical Guide Post-Jackson' by Andrew Mckie
'A Practical Guide to Subtle Brain Injury Claims' by Pankaj Madan
'A Practical Guide to Costs in Personal Injury Cases' by Matthew Hoe
'The No Nonsense Solicitors' Practice: A Guide To Running Your Firm' by Bettina Brueggemann
'The Queen's Counsel Lawyer's Omnibus: 20 Years of Cartoons from The Times 1993-2013' by Alex Steuart Williams

These books and more are available to order online direct from the publisher at www.lawbriefpublishing.com, where you can also read free sample chapters. For any queries, contact us on 0844 587 2383 or mail@lawbriefpublishing.com.

Our books are also usually in stock at www.amazon.co.uk with free next day delivery for Prime members, and at good legal bookshops such as Wildy & Sons.

We are regularly launching new books in our series of practical day-to-day practitioners' guides. Visit our website and join our free newsletter to be kept informed and to receive special offers, free chapters, etc.

You can also follow us on Twitter at www.twitter.com/lawbriefpub.